JENNIFER ACKERMAN

The Genius of Birds

PENGUIN BOOKS

PENGUIN BOOKS
An imprint of Penguin Random House LLC
375 Hudson Street
New York, New York 10014
penguin.com

First published in the United States of America by Penguin Press,
an imprint of Penguin Random House LLC, 2016
Published in Penguin Books 2017

ISBN 9781594205217 (hc.)
ISBN 9780399563126 (pbk.)

Printed in the United States of America
10 9 8 7 6 5 4 3 2 1

Designed by Michelle McMillian

Illustrations by John Burgoyne

FOR KARL (1955–2016), WITH ALL MY LOVE

Contents

The Genius of Birds

THE GENIUS OF BIRDS

For a long time, the knock on birds was that they're stupid. Beady eyed and nut brained. Reptiles with wings. Pigeon heads. Turkeys. They fly into windows, peck at their reflections, buzz into power lines, blunder into extinction.

Our language reflects our disrespect. Something worthless or unappealing is "for the birds." An ineffectual politician is a "lame duck." To "lay an egg" is to flub a performance. To be "henpecked" is to be harassed with persistent nagging. "Eating crow" is eating humble pie. The expression "bird brain," for a stupid, foolish, or scatterbrained person, entered the English language in the early 1920s because people thought of birds as mere flying, pecking automatons, with brains so small they had no capacity for thought at all.

That view is a gone goose. In the past two decades or so, from fields and laboratories around the world have flowed examples of bird species capable of mental feats comparable to those found in primates. There's a kind of bird that creates colorful designs out of berries, bits of glass, and blossoms to attract females, and another kind that hides up to thirty-three thousand seeds scattered over dozens of square miles and remembers where it put them months later. There's a species that solves a classic puzzle at nearly the same pace as a five-year-old child, and one that's an expert at picking locks. There are birds that can count and do simple math, make their own tools, move to the beat of music, comprehend basic principles of physics, remember the past, and plan for the future.

In the past, other animals have gotten all the publicity for their near-human cleverness. Chimps make stick spears to hunt smaller primates and dolphins communicate in a complex system of whistles and clicks. Great apes console one another and elephants mourn the loss of their own.

Now birds have joined the party. A flood of new research has overturned the old views, and people are finally starting to accept that birds are far more intelligent than we ever imagined—in some ways closer to our primate relatives than to their reptilian ones.

Beginning in the 1980s, the charming and cunning African grey parrot named Alex partnered with scientist Irene Pepperberg to show the world that some birds appear to have intellectual abilities rivaling those of primates. Before Alex died suddenly at the age of thirty-one (half his expected life span), he had mastered a vocabulary of hundreds of English labels for objects, colors, and shapes. He understood the categories of same and different in number, color, and shape. He could look at a tray holding an array of objects of various colors and materials and say how many there were of a certain type. "How many green keys?" Pepperberg would ask, displaying several green and orange keys and corks. Eight out of ten times, Alex got it right. He could use numbers to answer questions about addition. Among his greatest triumphs, says Pepperberg, were his knowledge of abstract concepts, including a zerolike concept; his capacity to figure out the meaning of a number label from its position in the number line; and his ability to sound out words the way a child does: "N-U-T." Until Alex, we thought we were alone in our use of words, or almost alone. Alex could not only comprehend words, he could use them to talk back with cogency, intelligence, and perhaps even feeling. His final words to Pepperberg as she put him back in his cage the night before he died were his daily refrain: "You be good, see you tomorrow. I love you."

In the 1990s, reports began to roll in from New Caledonia, a small island in the South Pacific, of crows that fashion their own tools in the wild and appear to transmit local styles of toolmaking from one generation to the next—a feat reminiscent of human culture and proof that sophisticated tool skills do not require a primate brain.

When scientists presented these crows with puzzles to test their problem-solving abilities, the birds astonished them with their crafty solutions. In 2002, Alex Kacelnik and his colleagues at Oxford University "asked" a captive New Caledonian crow named Betty, "Can you get the food that's out of reach in a little bucket at the bottom of this tube?" Betty blew away the experimenters by spontaneously bending a piece of wire into a hook tool to pull up the little bucket.

Among the published studies tumbling from scientific journals are some with titles that lift the brows: "Have we met before? Pigeons recognize familiar human faces"; "The syntax of gargles in the chickadee"; "Language discrimination by Java sparrows"; "Chicks like consonant music"; "Personality differences explain leadership in barnacle geese"; and "Pigeons on par with primates in numerical competence."

BIRD BRAIN: The slur came from the belief that birds had brains so diminutive they had to be devoted only to instinctual behavior. The avian brain had no cortex like ours, where all the "smart" stuff happens. Birds had minimal noggins for good reason, we thought: to allow for airborne ways; to defy gravity; to hover, arabesque, dive, soar for days on end, migrate thousands of miles, and maneuver in tight spaces. For their mastery of air, it seemed, birds paid a heavy cognitive penalty.

A closer look has taught us otherwise. Birds do indeed have brains very different from our own—and no wonder. Humans and birds have been evolving independently for a very long time, since our last common ancestor more than 300 million years ago. But some birds, in fact, have relatively large brains for their body size, just as we do. Moreover, when it comes to brainpower, size seems to matter less than the number of neurons, where they're located, and how they're connected. And some bird brains, it turns out, pack very high numbers of neurons where it counts, with densities akin to those found in primates, and links and connections much like ours. This may go a long way toward explaining why certain birds have such sophisticated cognitive abilities.

Like our brains, the brains of birds are lateralized; they have "sides" that process different kinds of information. They also have the ability to replace old brain cells with new ones just when they're needed most. And although avian brains are organized in an entirely different way from our brains, they share similar genes and neural circuits, and are capable of feats of quite extraordinary mental power. To wit: Magpies can recognize their own image in a mirror, a grasp of "self" once thought limited to humans, great apes, elephants, and dolphins and linked to highly developed social understanding. Western scrub jays use Machiavellian tactics to hide their food caches from other jays—but only if they've stolen food themselves. These birds seem to have a rudimentary ability to know what other birds are "thinking" and, perhaps, to grasp their perspective. They can also remember what kind of food they buried in a particular place— and when—so they can retrieve the morsel before it spoils. This ability to remember the what, where, and when of an event, called episodic memory, suggests to some scientists the possibility that these jays may be able to travel back into the past in their own minds—a key component of the kind of mental time travel once vaunted as uniquely human.

News has arrived that songbirds learn their songs the way we learn languages and pass these tunes along in rich cultural traditions that began tens of millions of years ago, when our primate ancestors were still scuttling about on all fours.

Some birds are born Euclideans, capable of using geometric clues and landmarks to orient themselves in three-dimensional space, navigate through unknown territory, and locate hidden treasures. Others are born accountants. In 2015 researchers found that newborn chicks spatially "map" numbers from left to right, as most humans do (left means less; right means more). This suggests that birds share with us a left-to-right orientation system—a cognitive strategy that underlies our human capacity for higher mathematics. Baby birds can also understand proportion and can learn to choose a target from an array of objects on the basis of its ordinal position (third, eighth, ninth). They can do simple arithmetic, as well, such as addition and subtraction.

Bird brains may be little, but it's plain they punch well above their weight.

BIRDS HAVE NEVER SEEMED dumb to me. In fact, few other creatures appear so alert, so alive in fiber and faculty, so endowed with perpetual oomph. Sure, I've heard the story of the raven attempting to crack open a Ping-Pong ball, presumably to get at an egglike morsel within. A friend of mine, while vacationing in Switzerland, watched a peacock try to fan its broad tail during a mistral. It toppled over, stood upright again, fanned again, and tipped over again, six or seven times in a row. Each spring the robins nesting in our cherry tree attack the side mirror of our car as if it were a rival, pecking furiously at their own reflections while streaking the door with guano.

But who among us hasn't been toppled by our own vanity or made an enemy of our own image?

I've watched birds most of my life and have always admired their pluck and focus and the taut, quick vitality that seems almost too much for their tiny bodies to contain. As Louis Halle once wrote, "A man would be worn out in short order by such intensity of living." The common species I saw around my old neighborhood appeared to negotiate the world with brisk curiosity and aplomb. The American crows striding around our garbage cans with a prince's proprietary air looked like highly resourceful creatures. I once watched a crow stack two crackers in the middle of a road before flying off to a safe spot to devour his collected booty.

One year, an eastern screech owl roosted in a box on a maple tree just a few yards from my kitchen window. In the daylight hours, the owl slept, only its round head showing, perfectly framed in the round hole facing the window. But at night, the owl was gone from the box, off hunting in the night. As the dawn light rose, there were signs of his brilliant success—the wing of a mourning dove or songbird hanging from the hole of the box, twitching, twitching, before it was yanked inside.

Even the red knots I encountered on the beaches of Delaware Bay, not the mentally swiftest of birds, seemed to know where to be—and when—to catch the rich feast of eggs laid by horseshoe crabs each full moon in spring. What calendar of sky drew these birds northward and told them where to go?

I LEARNED ABOUT BIRDS from a pair of Bills. The first was my father, Bill Gorham, who began taking me birdwatching near our home in Washington, D.C., when I was seven or eight. It was the Beltway version of a Swedish *gökotta*—the act of rising early to appreciate nature—and it was one of the palpable joys of my childhood. On early weekend mornings in spring we left the house in the dark and headed to the woods along the Potomac River to catch the dawn chorus, that mysterious moment when birds sing with a thousand voices in "A Music numerous as space— / But neighboring as Noon," as Emily Dickinson wrote.

My father learned about birds as a Boy Scout from a nearly blind man named Apollo Taleporos. The old man relied on his ears alone to pick up species. Parula warbler. Yellow-rumped warbler. Towhee. "The birds are there!" he would call out to the boys. "Go find them!" My father got very good at identifying birds by their calls—the melodious flutelike song of the wood thrush, the soft *whichity, whichity* of a common yellowthroat, or the clear whistling call of a white-throated sparrow.

As my father and I wandered through the woods in late starlight, I would listen to the husky song of a Carolina wren and wonder what, if anything, those birds were saying, and how they learned their songs. Once, I encountered a young white-crowned sparrow apparently engaged in song practice. There he was, perched invisibly somewhere in a low branch of a cedar tree, softly running through his whistles and trills, getting them wrong, and then going back over them quietly and persistently until he delivered the final run of his kind. This sparrow, I later learned, gleans his songs not from his own father but from birds in his natal environment, that very neighborhood of woods and rivers where my father and I rambled—a place with its own dialect passed down through the generations.

The other Bill I met at the Sussex Bird Club when I lived in Lewes, Delaware. Bill Frech was up and out of the house every morning at five A.M. for four or five hours of watching shorebirds and those little brown jobs, or LBJs, common in the woods and fields around Lewes. A patient, devoted, and inexhaustible observer, he kept meticulous notes on what birds he saw, where, and when, which ended up at the Delmarva Ornithological Society as part of the state's official bird records. This Bill was nearly deaf, but he was a wizard at identifying birds visually, by their so-called GISS, their general impression, size, and shape. He showed me how to spot a goldfinch high on the wing by its dipping flight and how to tell shorebirds apart by noting their personality, behavior, and gestalt, just as one recognizes friends from a distance by their overall manner and gait. He taught me the difference between casual "birdwatching" and the more intense, focused "birding," and urged me to go beyond identifying birds to noting their actions and behavior.

The birds I observed on those excursions and others seemed to know what they were doing. Like the black-billed cuckoo a friend saw perched just above a nest of tent caterpillars: The cuckoo waited as the caterpillars climbed out of the nest to scale the tree, then plucked them off one at a time, like sushi from a conveyor belt.

Still, I never imagined that the magpies and jays, the chickadees and herons, I admired so much for their feathers and flight, their songs and calls, might have mental abilities that match—even exceed—those in my primate tribe.

How can creatures with a nut-sized brain perform such sophisticated mental feats? What has shaped their intelligence? Is it the same or different from ours? What, if anything, do their little brains have to tell us about our big ones?

———

INTELLIGENCE IS a slippery concept, even in our own species, tricky to define and tricky to measure. One psychologist describes it as "the capacity to learn or to profit by experience." And another, as "the capacity to acquire capacity"—the same sort of circular definition offered up by Harvard psychologist Edwin Boring: "Intelligence is what is measured by intelligence tests." As Robert Sternberg, a former dean at Tufts University, once quipped, "There seem to be almost as many definitions of intelligence as . . . experts asked to define it."

In judging the overall intelligence of animals, scientists may look at how successful they are at surviving and reproducing in many different environments. By this measure, birds trump nearly all vertebrates, including fish, amphibians, reptiles, and mammals. They are the one form of wildlife visible nearly everywhere. They live in every part of the globe, from the equator to the poles, from the lowest deserts to the highest peaks, in virtually every habitat, on land, sea, and in bodies of freshwater. In biological terms, they have a very big ecological niche.

As a class, birds have been around for more than 100 million years. They are one of nature's great success stories, inventing new strategies for survival, their own distinctive brands of ingenuity that, in some respects at least, seem to far outpace our own.

Somewhere in the mists of deep time lived the überbird, the common ancestor of all birds, from hummingbird to heron. Now there are some 10,400 different bird species—more than double the number of mammal species: thick-knees and lapwings, kakapos and kites, hornbills and shoebills, chukars and chachalacas. In the late 1990s, when scientists estimated the total number of wild birds on the planet, they came up with 200 to 400 billion individual birds. That's roughly 30 to 60 live birds per person. To say that humans are more successful or advanced really depends on how you define those terms. After all, evolution isn't about advancement; it's about survival. It's about learning to solve the problems of your environment, something birds have done surpassingly well for a long, long time. Which to my mind makes it all the more surprising that many of us—even those of us who love them—have found it hard to swallow the idea that birds may be bright in ways we can't imagine.

Perhaps it's because they're so unlike people that it's difficult for us to fully appreciate their mental capabilities. Birds are dinosaurs, descended from the lucky, flexible few that survived whatever cataclysm did in their cousins. We are mammals, related to the timid, diminutive shrew-like creatures that emerged from the dinosaurs' shadows only after most of those beasts died off. While our mammal relatives were busy growing, birds, by the same process of natural selection, were busy shrinking. While we were learning to stand up and walk on two feet, they were perfecting lightness and flight. While our neurons were sorting themselves into cortical layers to generate complex behavior, birds were devising another neural architecture altogether, different from a mammal's but—in some ways, at least—equally sophisticated. They, like us, were figuring out how the world works, and all the while, evolution was fine-tuning and sculpting their brains, giving their minds the magnificent powers they have today.

BIRDS LEARN. They solve new problems and invent novel solutions to old ones. They make and use tools. They count. They copy behaviors from one another. They remember where they put things.

Even when their mental powers don't quite match or mirror our own complex thinking, they often contain the seeds of it—insight, for instance, one of our big-ticket cognitive abilities, which has been defined as the sudden emergence of a complete solution without trial-and-error learning. It often involves mental simulation of a problem and a kind of "aha!" moment when the solution becomes apparent in a flash of understanding. Whether birds have actual insight remains to be determined, but certain species seem to understand cause and effect—one of the building blocks of insight. The same is true for "theory of mind," a nuanced understanding of what another individual knows or thinks. Whether birds possess this full-blown ability is debatable, but members of certain species seem to be able to take the perspective of another bird or sense its needs, necessary components of theory of mind. Some scientists call these building blocks or stepping-stones the signatures of cognition and believe they may be the precursors to such highly complex human cognitive abilities as reasoning and planning, empathy, insight, and metacognition—awareness of one's own thought processes.

OF COURSE, these are all human yardsticks of intelligence. We can't help but measure other minds against our own. But birds also possess ways of knowing beyond our ken, which we can't easily dismiss as merely instinctual or hardwired.

What kind of intelligence allows a bird to anticipate the arrival of a distant storm? Or find its way to a place it has never been before, though it may be thousands of miles away? Or precisely imitate the complex songs of hundreds of other species? Or hide tens of thousands of seeds over hundreds of square miles and remember where it put them six months later? (I would flunk these sorts of intelligence tests as readily as birds might fail mine.)

Maybe *genius* is a better word. The term comes from the same root as *gene*, derived from the Latin word for "attendant spirit present from one's birth, innate ability or inclination." Later, *genius* came to mean natural ability, and finally (thanks to the essay "Genius" by Joseph Addison in 1711) to denote exceptional talent, natural or learned.

More recently, *genius* has been defined as "nothing more nor less than doing well what anyone can do badly." It's a mental skill that's exceptional compared with others, either of your kind or another kind. Pigeons have a genius for navigation that far, far exceeds our own. Mockingbirds and thrashers can learn and remember hundreds more songs than most of their fellow songbird species. Scrub jays and nutcrackers have memories for where they put things that make our capacity look meager.

IN THIS BOOK, *genius* is defined as the knack for knowing what you're doing—for "catching on" to your surroundings, making sense of things, and figuring out how to solve your problems. In other words, it's a flair for meeting environmental and social challenges with acumen and flexibility, which many birds seem to possess in abundance. Often this involves doing something innovative, something new—taking advantage of a new food source, for instance, or learning how to exploit it. The classic example of this was demonstrated years ago by tits in the United Kingdom. Both great tits and blue tits picked up the knack of opening the cardboard caps of milk bottles delivered to people's doorsteps in the morning to get at the rich cream on top. (Birds can't digest the carbohydrates in milk, only the lipids.) The tits first learned the trick in 1921 in the town of Swaythling; by 1949, the behavior had been noted in hundreds of localities throughout England, Wales, and Ireland. The technique had apparently spread by one bird copying another—an impressive show of social learning.

THE MISGUIDED USE OF "bird brain" as a slur has finally come home to roost. One by one, the bellwether differences between birds and our closest primate relatives seem to be falling away—toolmaking, culture, reasoning, the ability to remember the past and think about the future, to adopt another's perspective, to learn from one another. Many of our cherished forms of intellect—whether in whole or parts—appear to have evolved in birds quite separately and artfully right alongside our own.

How can this be? How can creatures separated by a 300-million-year gulf of evolution have similar cognitive strategies, skills, and abilities?

For one thing, we share more biology with birds than one might think. Nature is a master of bricolage, hanging on to biological bits that are useful and modifying them for new purposes. Many of the changes that separate us from other creatures have arisen not through the evolution of new genes or cells but through subtle shifts in how existing ones are used. This shared biology is what makes it possible to use other organisms as model systems to understand our own brains and behavior—to study learning in the giant sea snail *Aplysia*, anxiety in zebra fish, obsessive-compulsive disorder in border collies.

We also share with birds similar ways of meeting nature's challenges, which we've arrived at through very different evolutionary paths. It's called convergent evolution, and it's rampant in the natural world. The convergent shape of wings in birds, bats, and the reptiles known as pterosaurs results from the problems posed by flight. To meet the challenges of filter feeding, creatures as far apart on the tree of life as baleen whales and flamingos show striking parallels in behavior, body form (large tongues and hairy tissues known as lamellae), even body orientation during feeding. As evolutionary biologist John Endler points out, "Again and again, in totally unrelated groups, we find many instances of convergence in form, appearance, anatomy, behavior and other aspects. So why not in cognition, too?"

That both humans and certain species of birds have evolved brains large for their body size almost certainly represents convergent evolution. Likewise, the evolution of the same patterns of brain activity during sleep. And similarly, the evolution of analogous brain circuits and processes for learning song and speech. Darwin called birdsong "the nearest analogy to language." He was right. The parallels are eerie. Especially when you consider the evolutionary distance between humans and birds. A group of two hundred scientists from eighty different labs recently offered a window on these parallels when they sequenced the genomes of forty-eight birds. Their results, published in 2014, revealed startlingly similar gene activity in the brains of humans learning to speak and birds learning to sing, suggesting that there may be a kind of core pattern of gene expression for learning shared by birds and humans alike and arrived at through convergent evolution.

For all these reasons, birds are turning out to be wonderful animal models for understanding how our brains learn and remember, how we create language, what mental processes might underlie our problem solving, and how we locate ourselves in space and in social groupings. The circuits in the bird brain that control social behavior are much like the circuits in our own brains, it turns out, run by similar genes and chemicals. By investigating the neurochemistry of a bird's social nature, we stand to learn something about our own. Likewise, if we can grasp what's going on in a bird's brain as it masters a melody, we might get a better handle on how our own brains learn language, why it gets harder to master a new language over time, and maybe even how speech evolved in the first place. If we can understand why two animals so distantly related converged on the same pattern of brain activity during sleep, we might solve one of nature's great mysteries—the purpose of sleep.

THIS BOOK IS a quest to understand the different sorts of genius that have made birds so successful—and how they came about. It's a journey of sorts, venturing as far afield as Barbados and Borneo, as near as my own backyard. (You don't need to travel to exotic locations or see exotic species to witness intelligence in birds. It's everywhere around you, at your bird feeders, in local parks, city streets, and country skies.) It's also a voyage into the brains of birds, right down to the cells and molecules that drive their thinking and, sometimes, ours.

Each chapter tells the story of birds with extraordinary abilities or skills—technical, social, musical, artistic, spatial, inventive, adaptive. A few are exotic species; others, more common. You'll see members of the supremely clever corvid and parrot families appear and reappear throughout these pages, but also the sparrow and the finch, the pigeon and the chickadee. I'm interested in the everyman of the bird world as well as the Einsteins. I might have chosen other species as my stars, but I chose these for a simple reason: They have great stories to tell, stories that illuminate what might be going on in the mind of a bird as it solves the problems around it—and also, perhaps, give us some perspective on what is going on in our own minds. All these birds stretch our thinking about what it means to be intelligent.

The final chapter focuses on the adaptive brilliance of certain birds. Only a relative few possess this genius. Changes in the environment—especially those induced by humans—throw a wrench into the lives of many birds and disrupt their keen ways of knowing. A recent report from Audubon tells us that half of the bird species in North America—from whip-poor-will to white-tailed kite, common loon to shoveler, piping plover to dusky grouse—are likely to go extinct in the next half century or so for one reason: because they can't adapt to the rapid pace of human-induced change on our planet. Which birds will survive and why? In what ways are we humans an evolutionary force selecting for a certain kind of bird and bird intelligence?

SCIENTISTS ARE COMING at these puzzles from many different angles. Some are lifting the hood on the bird brain, using modern techniques to see what's going on in a bird's neural circuits when it recognizes a human face or to listen to individual brain cells as a songbird learns its song or to compare the neurochemicals in birds that are social butterflies with those that are loners. Some are sequencing and comparing bird genomes to pinpoint genes involved in complex behaviors such as learning. Others are strapping tiny geolocator backpacks on the backs of migratory birds to probe their journeys and their mapping minds. They're watching, tagging, measuring, conducting tireless observations, carefully preparing experiments at great length, some of which ultimately fail and must be reconfigured because their subjects are too wary or ornery. In short, these scientists are exploring the brains and behavior of birds in extraordinary, difficult—even heroic—ways.

But in this book, the birds themselves are the heroes of their own stories. My hope is that by the time you finish these pages, the chickadee and the crow, the mockingbird and the sparrow, will look a little different to you. More like the bright fellow sojourners they are—enterprising, inventive, cunning, playful, shrewd individuals that sing to one another in "accents," make complex navigational decisions without asking for directions, remember where they put things using landmarks and geometry, steal money, steal food, and understand the mental state of another individual.

Clearly there's more than one way to wire a clever brain.

one

FROM DODO TO CROW

Taking the Measure of a Bird Mind

The woods are cool, dark, and mostly quiet except for the occasional birdcall from somewhere in the thick canopy above, a patchwork of emerald, lichen, avocado, and a dark, coppery, almost iridescent green. This is typical mountainous rainforest on the island of New Caledonia, a remote tropical finger of land in the southwest Pacific, halfway between Australia and Fiji. The Parc des Grandes Fougères is named for the giant tree ferns that grow to seven stories and give this forest a truly primeval feel. The trail I'm following climbs for a while, then dips down toward a stream, where the birdsongs and calls grow louder.

I have come to this island to see what is arguably the world's smartest bird, the New Caledonian crow (*Corvus moneduloides*), a member of the common but uncommonly intelligent corvid family. It's a bird species made famous by Betty, that crow who some years ago appeared to spontaneously bend a piece of wire into a hook to fetch some hard-to-get food. And more recently by a wizard of a bird nicknamed "007," who became a star in 2014 when his speedy solution to a challenging eight-step puzzle was captured on film by the BBC.

The puzzle was created by Alex Taylor, a senior lecturer at the University of Auckland in New Zealand. It consisted of eight separate stages, each made of various special chambers and "toolboxes" containing sticks and stones, all laid out on a tabletop. The bird 007 had seen the different parts of the puzzle but never in this particular configuration. To win the cube of meat in the final special food chamber, he had to work out the puzzle's steps in the proper order.

In the video, a dark, handsome (aptly named) bird flies into view, settles on a perch, and takes a few moments to scope out the situation. Then he flutters up to a branch that holds a stick hanging from a string—the first step in the puzzle. He pulls up the stick, one string length at a time, until he can grasp the stick with his beak. He drops down from perch to tabletop, hops over to the food chamber, and uses the stick to poke into the food chamber's deep horizontal hole to try to collect the treat. But the stick is too short, so he uses it instead to retrieve three stones from three separate boxes. These he drops one at a time into a hole at the top of a chamber containing a longer stick balanced on a seesaw. The weight of the three stones tilts the seesaw inside the box, releasing the long stick, which the bird takes back to the food chamber to lever out his meat.

It's an astonishing process, and it takes the bird all of two and a half minutes to complete. The really clever part is this: The eight-stage puzzle requires understanding that a tool can be used not just to get at food directly but to get at another tool that will help access the food. Spontaneously aiming a tool at an object that's not food but deemed useful to secure another tool—known as metatool use—has been seen only in humans and great apes. "It suggests that the crows have an abstract understanding of what a tool does," says Taylor. The task also demands working memory, the ability to keep in mind facts or thoughts and manipulate them for a short time—a few seconds or so—while solving a problem. Working memory is what allows us to remember what we're looking for when we scan a bookshelf for a particular title or recall a phone number while we pull out a piece of paper to write it down. It's a vital component of intelligence, which this crow seems to possess in spades.

FROM SOMEWHERE ALONG the stream, I catch the *wak, wak* of one New Caledonian crow, maybe two calling to each other—not unlike the *caw, caw* of an American crow, only played in reverse. Birds are so often encountered this way, as disembodied voices. The low, mournful *woo, woo, woo* in the distance might be the little green foghorn of a cloven-feathered dove, an exotic harlequin of a bird with bands of white and dark green across its wings and rump. But the canopy here is so thick I can't make out any birds at all.

The sun passes behind a cloud and the forest darkens. Suddenly, from the understory, I hear a weird sibilant hiss. I peer into the glade. The hiss moves closer. Then out of the green gloom I see a large pale bird running toward me like a spirit loosened from the ground, a hybrid of bird and ghost. It's heronlike, knee high, with a cockatoo crest, but smoky gray: the flightless kagu (*Rhynochetos jubatus*), sole representative of its family and among the hundred rarest birds on earth.

I had been looking for a supremely clever bird, common in these parts. And here I had stumbled on a most uncommon bird, one that appeared to be . . . well . . . a few crayons short of the box. The kagu is near extinction; its population numbers in the hundreds. And no wonder, I thought. A bird that runs *toward* a possible predator?

In a way, the kagu seems a bookend to the crow, a delegate of the dim end of the intelligence spectrum. How could this creature be in the same phylogenetic class as the cunning crows? Both birds occupy the same remote island. Are New Caledonian crows evolutionary anomalies, hyper-intelligent deviants that have advanced far beyond their feathered peers? Or are they simply at the high end of the bird genius continuum? By the same token, is the kagu really such a dodo?

Clearly all birds are not equally bright or able in all ways—at least by current accounting. Pigeons, for instance, don't do well on tasks that require them to abstract a general rule to solve a suite of similar problems, a skill easily learned by crows. But the lowly pigeon is a wizard in other ways: It can remember hundreds of different objects for long periods of time, discriminate between different painting styles, and figure out where it's going, even when displaced from familiar territory by hundreds of miles. Shorebirds such as plovers, sanderlings, and sandpipers show no evidence of "insight learning," that grasp of relationships that permits birds like the New Caledonian crow to use tools or to operate man-made devices that reward their ingenuity with food. But one shorebird, the piping plover, is a master of theatrics, capable of diverting predators from their shallow, exposed nests with a feigned "injured wing" display.

What makes one bird smarter than another? How do you measure a bird's intelligence anyway?

TO EXPLORE THESE QUESTIONS, I traveled to a spot half a world away from New Caledonia: the Caribbean island of Barbados, where more than a decade ago Louis Lefebvre invented the first scale of intelligence for birds.

A biologist and comparative psychologist at McGill University, Lefebvre has spent his career inquiring into the nature of the bird mind and how to take its measure. One winter not long ago, I went to see him and his birds at the Bellairs Research Institute, a hodgepodge of four small buildings near Holetown on the west coast of Barbados, where he conducts his studies. The institute is a small estate that was bequeathed to McGill in 1954 by Commander Carlyon Bellairs, a British naval officer and politician, for use as a marine research station. These days, few researchers use the place except Lefebvre and his team. It was February, the middle of the dry season in Barbados, but monsoonlike downpours passed through frequently, drenching the institute's courtyard and forming pools of water in the dips and depressions of the terrace at Seabourne, the residential building hard by the Caribbean Sea, where Lefebvre stays when he's conducting his research.

Sixtysomething, with an easy smile and a shock of curly grayish-black hair, Lefebvre trained under the evolutionary biologist Richard Dawkins. He first studied grooming in animals, an innate, "programmed" behavior; now he aims to understand the more complex behavior of birds—how they think, learn, and innovate—using the weedy bird species in his own Bajan backyard.

Unlike New Caledonia, Barbados is not a spot to goose your life list. Measured against the lush diversity found in most tropics, the island disappoints. It's marked by a distinctly "depauperate avifauna," as the experts say, home to only thirty native breeding species and seven introduced species. This is partly due to its physical nature. A tiny, low stack of young coral limestone east of the main chain of Lesser Antillean islands, Barbados is too flat for rainforest and too porous for creeks and marshes. Moreover, in the past few centuries, the island's natural fields, forests, and scrubland have been planted over with sugarcane. These days Barbados is heavily developed with towns and facilities for tourism. From the open windows of the painted buses shuttling back and forth between hotels and beaches come calypso tunes. Flourishing here are the few bird species that advance rather than recede in the face of this human expansion. For a birder set on spotting rare species such as the kagu, Barbados is a wasteland. But if you're keen on watching birds do smart, beguiling things, it's a paradise.

"The tameness of the birds here makes it easy to do experiments," says Lefebvre. The wide stone terrace right in front of his apartment, for instance, is a sort of casual lab, where zenaida doves—the pigeons of Barbados—and Carib grackles mull around, waiting for action. The grackles (fittingly named *Quiscalus lugubris*) are glossy black with a bright yellow eye, smaller than the American boat-tailed variety and more compact. They know that Lefebvre is the "pellet and water guy," as he says, and are pacing about the terrace like impatient clergymen, waiting for him to deliver. He empties a pan of water on the terrace, creating a little lake, and throws out some tough pellets of dog food on the dry portion. The grackles seize a pellet in their bills and strut over to the puddle, ceremoniously and delicately dip the pellet in the water, then flap off to eat the softened food.

More than twenty-five species of birds dunk food in the wild for one reason or another—to wash soiled or toxic items, to soften hard or dry ones, or to smooth the fur or feather of hard-to-swallow prey (like the Torresian crow, seen dunking a dead sparrow). "It's proto-tool behavior, a kind of food-processing," Lefebvre explains. The dunking makes the pellet easier to eat. "Once I presoaked the pellets, and they stopped dunking. They walked over to the puddle, but didn't actually dunk. So they know what they're doing."

In Carib grackles, dunking is a relatively rare behavior because it's risky. "Our studies show that 80 to 90 percent of these grackles are capable of the behavior, but they'll only do it if the circumstances are right," says Lefebvre, "the quality of food, the social conditions, who's around to compete or steal." The longer food-handling time increases the risk of theft from other grackles that scrounge or pilfer. "Theft is a major cost of dunking," he explains. Up to 15 percent of items are stolen by competitors. "There's a cost/benefit ratio, and the birds are smart enough to determine whether it's worth it." That seems like intelligent behavior by any measure.

ANIMAL SCIENTISTS TEND to shy away from the term *intelligence* because of the human connotations it carries, Lefebvre tells me. In his *History of Animals*, Aristotle wrote that animals carry elements of our "human qualities and attitudes," such as "fierceness, mildness or crosstemper, courage or timidity, fear or confidence, high spirits or low cunning, and with regard to intelligence, something akin to sagacity." These days, however, suggest that a bird has anything like human intelligence, consciousness, or subjective feeling, and people might accuse you of anthropomorphizing, interpreting the behavior of a bird as if it were a human clothed in feathers. It's a natural human impulse to project our own experience on the nature of other creatures, but this can—and does—lead us astray. Birds, like humans, are kingdom *Animalia*; phylum *Chordata*; subphylum *Vertebrata*. There the common descent ends. Birds are class *Aves*; we are *Mammalia*. And in that branching is a mountain of biological difference.

But wouldn't it be a mistake to assume that because birds and their brains are fundamentally different from us and ours, there is nothing in common between our mental abilities and theirs? We call our species *Homo sapiens*, the sapient, to distinguish us from the rest of life. However, in *The Descent of Man*, Darwin argued that animals and humans differ in their mental powers only in degree, not in kind. For Darwin, even earthworms "show some degree of intelligence" in their manner of dragging pine needles and vegetable matter to plug up their burrows, protection from the proverbial "early bird." Tempting as it may be to interpret the behavior of other animals in terms of human mental processes, it's perhaps even more tempting to reject the possibility of kinship. It's what primatologist Frans de Waal calls "anthropodenial," blindness to the humanlike characteristics of other species. "Those who are in anthropodenial," says de Waal, "try to build a brick wall to separate humans from the rest of the animal kingdom."

IN ANY CASE, says Lefebvre, "you have to watch your vocabulary." He points to a study published recently on empathy in mice and one on mental time travel in birds, which raised both eyebrows and doubts. "I'm not questioning the experiments—they are sound, and they do not anthropomorphize," he says. "But perhaps we go too far in the words we use to describe what we think is going on."

Like Lefebvre, most scientists who study birds prefer the term *cognition* to *intelligence*. Animal cognition is generally defined as any mechanism by which an animal acquires, processes, stores, and uses information. It usually refers to the mechanisms involved in learning, memory, perception, and decision making. There are so-called higher and lower forms of cognition. For instance, insight, reasoning, and planning are considered high-level cognitive abilities. Lower-level cognitive skills include attention and motivation.

There's less consensus about what form cognition takes in a bird's mind. Some scientists suggest that birds possess distinct types of cognition—spatial, social, technical, and vocal—that don't necessarily dovetail. A bird can be smart spatially without being gifted in social problem solving. In this view, the brain is seen as a bundle of different specialized processors, or "modules," discrete zones adapted and dedicated to a particular purpose—like the circuitry for learning birdsong or for navigating through space. The information in each module is essentially "unavailable" to other modules. Lefebvre, on the other hand, argues for such a thing as general cognition—one all-purpose processor, untidily distributed, for problem solving in different domains—pointing out that if a bird ranks high on one cognitive measure, it tends to rank high on others. "When an animal is solving problems," he says, "different zones in the brain are likely involved in a network of interactions."

According to Lefebvre, some scientists in the modular camp are beginning to inch toward his view as studies uncover evidence that some birds may be using general cognitive mechanisms to solve different sorts of problems. For instance, he says, social intelligence in some birds appears to go hand in hand with spatial memory or episodic-like memory—the ability to remember what happened where and when.

There's a parallel discussion in human intelligence. Most psychologists and neuroscientists agree that there are different kinds of human intelligence—emotional, analytic, spatial, creative, practical, to name a few. But they still debate about whether the types are independent or correlated. In his theory of "multiple intelligences," Harvard psychologist Howard Gardner identifies eight different types of intelligence and suggests that they're independent. They are bodily, linguistic, musical, mathematical or logical, naturalistic (sensitivity to the natural world), spatial (knowing where you are relative to a fixed location), interpersonal (sensing and being in tune with others), and intrapersonal (understanding and controlling one's own emotions and thoughts)—a list with intriguing parallels in the bird world: Think of a hummingbird's acrobatic use of its own body or a plain-tailed wren's astonishing talent for musical duets or a pigeon's gift for knowing where it needs to go.

Other scientists argue for such a thing as general intelligence in humans, being all-around smart, a g factor, as it's known. A group of fifty-two researchers who assembled to tackle the matter some years ago agreed: "Intelligence is a very general capability that, among other things, involves the ability to reason, plan, solve problems, think abstractly, comprehend complex ideas, learn quickly and learn from experience."

IF DEFINING INTELLIGENCE in birds is problematic, measuring it is perhaps even harder. "The truth is, the development of a battery of tests to measure bird cognition is still in its infancy," says Lefebvre. There's no standard IQ test for birds. So scientists try to devise puzzles for birds that reveal their cognitive abilities, comparing the performance between species and also between individuals of the same species.

A nondescript little brown Bajan bird plays a key role in Lefebvre's recent investigations. While I sit writing up notes on the back porch of Lefebvre's apartment overlooking an azure sea, the little brown birds flit about in the branches of Australian casuarinas and mahogany trees nearby. Then they pop right onto the terrace railings. I stare at one perched within arm's reach. He flips around, cocks his head, and stares back at me.

Why the keen interest? he seems to ask.

Because you're notorious around here for your clever, thieving ways—and for your discovery of a new food source.

Loxigilla barbadensis: These bullfinches are the house sparrows of Barbados, says Lefebvre. Before there were screens installed in the building to protect against dengue fever, the bullfinches would fly through his apartment windows or doors open to the sea air to ravage bananas on the kitchen counter or make off with chunks of bread or cake. But their claim to fame is their discovery of a new food source in the outdoor restaurants lining the Caribbean Sea. Later, Lefebvre shows me the bird's special food trick. In a narrow alley between two shorefront clubs in Holetown is a stone wall edging a Palladian-style mansion by the sea. Lefebvre places a packet of sugar on a rock and then lines up four more along the wall. It takes only seconds for a bullfinch to find the treasure. It lands on the wall and investigates the little white paper square, flips it over, apparently inspecting for holes, then carries it up to a nearby tree branch. In thirty seconds it has poked through the paper and is feeding on the sugar, white crystals coating its little bill like milk around a child's mouth. It's a unique talent, not mastered by the handful of other weedy species that make this island their home. This bullfinch knows what it's doing. It's bold, brazen, and quick to explore new sources of food.

It was here in the land of the bullfinch that Lefebvre devised a scale of intelligence based on the idea that smart birds innovate. Like the bullfinch and those cream-skimming tits, they do new things. Birds with lesser brains are set in their ways and rarely invent, explore, or dip into the novel.

As it happens, the Bajan bullfinch has a dimmer doppelgänger on the island, the closely related black-faced grassquit (*Tiaris bicolor*), which provides an intriguing comparison. The two birds are alike in nearly every respect, save one. On the spectrum of intelligence, the bullfinch is a quick study; in comparison, its sister species is slow and plodding. The contrast between these two backyard species has offered Lefebvre a window on the nature of the bird mind.

"These two birds are virtual genetic twins with the same ancestor, having diverged probably only a couple of million years ago," Lefebvre explains. "Both live in the same environment. Both are territorial and share the same social system." The only difference is that the bullfinch is clever, fearless, and opportunistic; and the grassquit, skittish, deeply conservative, and afraid of nearly everything.

The evolutionary backstory of the bullfinch may be telling. When it arrived on Barbados, the species diverged from the colorful Lesser Antillean bullfinch. In that species, males and females are dimorphic in color, with females a plain brown and males sporting a sexually selected plumage of handsome black with a bright red throat. Here in Barbados, the bullfinches are monomorphic, both genders an equally humble brown.

"One explanation for this evolutionary shift is that Barbados didn't have the carotenoid-based foods that allowed the birds to produce reds and yellow for plumage," says Lefebvre. "But as it turns out, the bird's red plumage doesn't require carotenoids. It's possible that females are selecting for something other than plumage. Maybe they're going for males that seek innovative food sources—like sugar packets." In other words, maybe female bullfinches on Barbados like their males on the smart side.

"I don't know of any other pair of closely related species that are so similar and yet so different in their opportunism and foraging strategies," says Lefebvre. In a small expanse of woods and fields at Folkestone Marine Park, he offers an informal experiment to demonstrate his point. Several grassquits are visible, poking about in the grass thirty yards away, foraging on seed. A few other birds are off in the trees at some distance. Lefebvre throws out a handful of birdseed, then squats in the grass. The grackles are the first to notice. Within half a minute, they're gathering around in a noisy flock. Their squawks draw doves, more grackles, and squadrons of bullfinches. The grassquits have not budged. They just keep their heads down, closely attending only to their little plots of grass. Lefebvre lowers his voice to a whisper and says in a British accent: "A perfect result, as if staged, with David Attenborough hiding in the wings." And in an uncanny imitation of the famous naturalist: "This bird does *amazing* things . . ."

He stands up abruptly and points at the grassquits. "Zero opportunism there," he says. "They're attracted neither by the seed nor by the birds feeding on it. They're just not on the lookout for alternative food sources."

For fifteen years Lefebvre ignored grassquits because they seemed so . . . well . . . so *boring*. But now they present a perfect experimental pairing with the bullfinch because of their genetic closeness.

"Why is the grassquit the way it is?" wonders Lefebvre. "It has the same ancestral genotype as the bullfinch, lives in the same environment.

What makes it take such a totally different approach to food?" Why is one bird so much bolder, smarter, and more opportunistic than the other?

"Studies show that species that differ in feeding ecology also differ in learning ability—and in the brain structure underlying learning," says Lefebvre. So first up is an experiment presenting both birds with tasks to measure their basic cognitive abilities. It's a step toward linking the natural behavior the scientists see in the field with differences they can measure in the laboratory.

It's not an easy task. Just catching the grassquits is challenging. Lefebvre uses walk-in traps for bullfinches, but in twenty-five years of work here, he has never caught a grassquit in such a trap; the birds are far too wary. So the team uses mist nets to capture their subjects.

"Then the trick is to find something the grassquits will do," says Lefebvre. "They're so skittish that if an experimental apparatus is a little too weird, they just don't participate." In the field, one of Lefebvre's graduate students, Lima Kayello, has measured the speed at which the two species will feed from an open cup of seed. The bullfinches find the novel food source in about five seconds, she says. It takes the grassquits five days. "A yogurt top filled with seeds is just too odd for them," says Kayello.

For the cognitive experiments, Kayello presents each of the two species with something they've never seen: a small transparent cylinder of food with a removable lid. She measures how long it takes the birds to approach the apparatus, make contact with it, and then flip the lid and feed on the seed. There's variety of performance, even among the bullfinches. One bullfinch flits around the aviary for several minutes, then hangs batlike from the lowest perch for several more minutes, before finally venturing toward the apparatus and opening it. It takes him a total of eight minutes to solve the task. A second bird goes straight to the new gizmo and opens it almost immediately. "Good boy!" says Kayello. Time spent on trial: seven seconds.

Of the thirty bullfinches Kayello tested, twenty-four quickly completed the obstacle removal task. Not one of the fifteen grassquits even got close to the cylinder.

Some bullfinches, like that second one, seem able to figure out how to solve the problem quickly, with very few attempts. Is this an example of insight? Lefebvre doesn't think so. In a comparable study, his graduate student Sarah Overington examined every peck a grackle made in a similar problem-solving test. After scrutinizing hundreds of hours of videos, Overington observed that the birds had two types of pecks. The first type of peck was an attempt to get directly at the food; the second type was to the side, which made the lid move, giving the bird a cue to keep going with the pecking. Even a little visual or tactile feedback can direct the bird. "If it was insight," says Lefebvre, "you would expect a sudden solution to the problem, a kind of eureka!" This is more like trial-and-error learning, a "lesser" cognitive ability.

THE POINT IS, behaviors that seem extraordinary or intelligent may arise out of simple or reflexive processes.

One striking example of this is cluster flocking—birds or other creatures moving in apparent unison, sometimes in large numbers. I was once drawn to my yard by a cacophony of starlings beading our hackberry tree with black, twittering blossoms. Suddenly the shadow of a hawk passed over, and the blackbirds exploded upward, almost as one, and swirled away. I watched the whole shimmering sheet of them dark against the sky, wheeling, twisting, eddying in intricate movements with the cohesion of a single organism—an effective strategy for deterring a predator like a hawk or a falcon. The great naturalist Edmund Selous, who loved birds passionately and observed them with scientific fervor, attributed this flocking phenomenon to telepathic thought transference from one bird to the next. "They circle; now dense like a polished roof, now disseminated like the meshes of some vast all-heaven-sweeping net, now darkening, now flashing out a million rays of light . . . a madness in the sky," he wrote. "They must think collectively, all at the same time, or at least in streaks or patches—a square yard or so of an idea, a flash out of so many brains."

We've since learned that the spectacular collective behavior of flocking birds (and schooling fish, herding mammals, swarming insects, and human crowds) is self-organized, emerging from simple rules of interaction among individuals. Birds are not "transfusing thought," communicating telepathically with their flock members to act in unison, as Selous surmised. Instead, each bird is interacting with up to seven close neighbors, making individual movement decisions based on maintaining velocity and distance from fellow flock members and copying how sharply a neighbor turns, so that a group of, say, four hundred birds can veer in another direction in a little over half a second. What emerges is almost instantaneous ripples of movement in what appears to be one living curtain of bird.

IT'S A COMMON ASSUMPTION that apparently complex behavior must arise from complex thought processes. But the quick problem-solving abilities of the bullfinches and grackles in these basic cognition tests probably have more to do with paying attention to visual feedback and self-correcting along the way than instantly "figuring out" a solution.

In another test of cognition, Kayello tries to get the birds to unlearn what they've learned and "relearn" something new. She presents each bird with two cups, one yellow, one green, filled with edible seed, and lets the bird pick one to eat from to find out its color preference. She then replaces the edible seed in that colored cup with inedible seed that is glued to the bottom of the cup. She measures how long it takes for each bird to switch over from the cup with its preferred color (but now containing inedible seeds) to the other, nonpreferred colored cup (with edible seeds). When that's done, she again switches the colors that identify the edible and inedible seeds.

This technique, called reversal learning, is often used as a basic measure of how rapidly a bird can switch its thinking and learn a new pattern. "It's an indicator of flexible thinking," Lefebvre explains. "For humans as well as birds. People with mental deficits or Alzheimer's disease are often tested with reversal learning tasks to check their flexibility of thinking."

There's no question—the bullfinches are quick studies. Most get the hang of the switching after only a few trials. The grassquits, on the other hand, take their time. They're slow, wary. But eventually they master the trick and end up making fewer wrong color choices than the bullfinches.

"Surprising," says Lefebvre, "but reassuring in a way: At least we found one test that grassquits do well. If one of the species you're using in your experiment fails every test you give it, the problem may be you, the researcher, not the animal. You may have failed to understand what is relevant to the way a bird sees the world."

THIS IS ONE WAY scientists attempt to measure a bird's intelligence— by testing its speed and success at solving problems in the laboratory. They try to design challenges similar to those a bird might encounter in its natural environment—say, the ability to remove obstacles, or to navigate around barriers to find hidden foods. They ask birds to open food containers by pushing levers, pulling strings, swinging caps aside. They measure how long it takes and how readily birds change tactics in attempting to solve the problem. ("If x doesn't work, try y.") They test for insight by trying to determine whether a bird's discovery of a solution is a sudden flash of understanding (eureka!) or gradual and more reflexive (trial and error).

It's tricky, however. In these kinds of lab tests, all sorts of variables may affect a bird's failure or success. The boldness or fear of an individual bird may affect its problem-solving performance. Birds that are faster at solving tasks may not be smarter; they may just be less hesitant to engage in a new task. So a test designed to measure cognitive ability may really be measuring fearlessness. Is the grassquit just a shier bird?

"Unfortunately it is extremely difficult to get a 'pure' measure of cognitive performance that is not affected by myriad other factors," says

Neeltje Boogert, a former student of Lefebvre, now a bird cognition researcher at the University of St Andrews. "Birds, just like humans, will differ in how motivated they are to solve a cognitive test, how stressed they are about the test situation, how distracted they are by their surroundings, and how much experience they have had with similar tests. A raging debate is ongoing in the field of behavioral ecology on how we should go about testing animal cognition; thus far no clear solutions have emerged."

SOME YEARS AGO, Lefebvre was struck by the possibility for another sort of measure, one that would gauge a bird's cognitive ability not in the laboratory but in the wild. The idea was sparked by chance during a walk on the beach in Barbados. "It was right after a violent storm," he says. "I was crossing the beach near the Hole, the lagoon in Holetown that overflows into the sea after heavy rains, and I noticed that several hundred guppies had gotten trapped in small ponds on a sandbar." As the stranded fish flipped from one pool to the next, Lefebvre saw gray kingbirds swoop down to pick off the fish, take them back to a tree, and hammer them against a branch before eating them.

Gray kingbirds are a kind of common West Indian flycatcher. They're well known for catching insects on the wing—but not fish. This was the first known observation of the birds applying their usual hunting skills to an entirely unusual prey.

Lefebvre wondered, "Why was the kingbird the one bird taking advantage of this splendid new food source?" Was it an especially intelligent or innovative species, like the tits that cracked the code of milk bottle caps to get at the cream?

Maybe a good way to measure bird cognition, Lefebvre thought, would be to look at these sorts of occurrences—birds doing unusual new things in the wild. It's a notion that was proposed three decades ago by Jane Goodall and her colleague Hans Kummer. The pair made a plea for measuring a wild animal's intelligence by looking at its ability to find solutions to problems in its natural setting. What's needed is an ecological rather than a laboratory measure of intelligence, they suggested. This can be found in an animal's ability to innovate in its own environment, "to find a solution to a novel problem, or a novel solution to an old one."

Lefebvre published his observation on the gray kingbirds in the notes section of the *Wilson Bulletin*, which issues reports of uncommon bird behavior by amateur birdwatchers and professionals alike. It occurred to him that collecting these sorts of anecdotal notes from ornithological journals might provide just the sort of ecological evidence Kummer and Goodall called for. Which kinds of birds are the most innovative in the wild?

"Experimental and observational studies of cognition are important," says Lefebvre, "but a taxonomic count like this would provide a unique opportunity and would avoid some of the pitfalls of animal intelligence studies"—for instance, the use of testing devices that are far removed from what an animal does in its natural environment.

Lefebvre scoured seventy-five years' worth of bird journals for reports featuring key words like "unusual," "novel," or "first-reported instance," and came up with more than twenty-three hundred examples from hundreds of different species. Some of these were daring discoveries of strange new foods: a roadrunner sitting on a roof next to a hummingbird feeder and picking off the hummers; a great skua in the Antarctic snuggling in among newborn seal pups and sipping milk from their lactating mother; herons wolfing down a rabbit or a muskrat; a pelican in London swallowing a pigeon; a gull ingesting a blue jay; or a normally insectivorous yellowhead in New Zealand seen for the first time eating bush lily fruits.

Other examples involved ingenious new ways of getting at food. There was the cowbird in South Africa using a twig to pick through cow dung. Several observers noted instances of green-backed herons using insects as bait, placing them delicately on the surface of the water to lure fish. A herring gull adapted its normal shell-dropping technique to nail a rabbit. Among the more inventive examples: bald eagles ice fishing in northern Arizona. The birds had discovered a cache of dead fathead minnows frozen under the surface of an ice-covered lake. They were seen chipping holes in the ice, then jumping up and down on the surface, using their body weight to push the minnows up through the holes. One of Lefebvre's favorites was the report of vultures in Zimbabwe that perched on barbed-wire fences near minefields during the war of liberation, waiting for gazelles and other grazers to wander in and detonate the explosives. It gave the birds a ready-made meal already pulverized. However, says Lefebvre, "occasionally a vulture got caught at its own game and was exploded by a mine."

Once the anecdotes were gathered, Lefebvre grouped them by bird family and calculated the innovation rates for each family. He also corrected his analyses to account for possible confounding variables, especially research effort—some species are simply more often observed, so they're more likely to be seen doing novel things.

"Honestly, initially I didn't think it would work," he says. Anecdotes are considered unscientific; they're "weak data," in the lingo. "If one anecdote is unscientific, how can two thousand anecdotes become science? But I accepted the data at face value. If there was slough in the database, it was probably randomly distributed across taxonomic groups, so it wouldn't affect the results. I've been waiting for something to come up that would invalidate the system, but nothing has."

What are the smartest birds according to Lefebvre's scale?

Corvids, no surprise—with ravens and crows as the clear outliers—along with parrots. Then came grackles, raptors (especially falcons and hawks), woodpeckers, hornbills, gulls, kingfishers, roadrunners, and herons. (Owls were excluded from the search because they are nocturnal and their innovations are rarely observed directly, but rather inferred from fecal evidence.) Also high on the totem pole were birds in the sparrow and tit families. Among those at the low end were quails, ostriches, bustards, turkeys, and nightjars.

Lefebvre then took his scale a step further: Did families of birds that showed a lot of innovative behaviors in the wild have bigger brains? In most cases, there was a correlation. Consider two birds weighing

320 grams: The American crow, with an innovation count of sixteen, has a brain of 7 grams, while a partridge, with one innovation, has a brain of only 1.9 grams. Or two smaller birds weighing 85 grams: the great spotted woodpecker, with an innovation rate of nine, has a brain weighing 2.7 grams, and the quail, with one innovation, only 0.73 gram.

When Lefebvre presented his findings at the annual meeting of the American Association for the Advancement of Science in 2005, the press picked up on the study, calling it the world's first comprehensive avian IQ index. Lefebvre found the IQ idea "a little cheap," he said. "But why not?"

The notion caught on, and Lefebvre ended up being quizzed closely by interested journalists. When one asked him to name the world's dumbest bird, Lefebvre answered, "That would be the emu." The next day's headlines read, CANADIAN RESEARCHER NAMES NATIONAL BIRD OF AUSTRALIA "WORLD'S MOST STUPID BIRD." (The emu and the kangaroo were selected as unofficial emblems of Australia to symbolize the forward movement of a nation, reflecting a common—and false—belief that neither animal can easily move backward.) This did not make Lefebvre popular in Australia. But his position was buoyed when he appeared on an Australian radio show and one caller related a story of being in the outback with aboriginals, who told him that if he lay down on his back and raised his foot, the emus would come to investigate, thinking he was one of them.

LEFEBVRE ACKNOWLEDGES THAT the size of a bird's brain or even the size of key parts of it is a relatively crude measure of intelligence. "After all, the little stint (a sandpiper) has a relatively large brain for its body size," he says, "and all it does is skitter back and forth with the waves ('can't get my knees wet, can't get my knees wet'), pecking at invertebrates."

We've known for a long time that a big brain is not necessarily a proxy for smarts. A cow has a brain one hundred times the size of a mouse's, but it isn't any smarter. And animals with minute brains have surprising mental abilities. Honeybees, with a brain weighing 1 milligram, can navigate landscapes on par with mammals, and fruit flies can learn socially from other fruit flies. The ratio of brain size to body size, called brain encephalization, seems to play into the picture, although how closely encephalization correlates with intelligence is still a matter of debate.

"It's not just about size—at least not in all animals," says Lefebvre. "When we're measuring brain volume, are we measuring information-processing capacity?" asks Lefebvre. "Probably not."

A BIRD'S ABILITY to innovate is now accepted as a measure of cognition by many scientists. But if brain size doesn't control a bird's tendency to innovate, what does? What distinguishes innovators from noninnovators? Is there a difference between the same-size brains of the bright bullfinch and the apparently duller grassquit?

"The problem is to get inside an animal's head," says Lefebvre. "Until now, the focus has all been on brain volume, whole, or of particular parts. But that's not really where it's happening. What's controlling innovation and cognitive ability is not size, but what's going on at the level of the neuron."

This brings to mind the advice that neuroscientist Eric Kandel, who won the Nobel Prize for his work on the physiological basis of memory storage in neurons, took from his mentor, Harry Grundfest. When Kandel was a young man, Grundfest told him, "Look, if you want to understand the brain you're going to have to take a reductionist approach, one cell at a time." Says Kandel. "He was so right"

Like many other bird cognition researchers, Lefebvre is now "going neural," hoping to show how learning and problem solving in birds is reflected in brain activity, in neurons and in the junctions between them, known as synapses. One neuron communicates with another at these links between the two cells. "I believe that whether an animal shows flexible, innovative behavior or not depends on events occurring here, at the synapses," says Lefebvre.

What makes a bird smart and inventive like a Bajan bullfinch or a

New Caledonian crow? Are black-faced grassquits and kagus really such simpletons?

"We're trying to come at these questions from different angles," says Lefebvre. "You have to begin in the field, with your boots on the ground, and really look closely at the species in question. If you want to understand birds, you have to know how they behave in the wild," he says. "Then you try to get inside their heads. So we're doing field observations of behaviors, comparing innovations per species, conducting experiments with captive birds, and now, looking for a way to connect what we see in the field with what we're learning about genes and cells in the lab."

This is the sort of ambitious science going on everywhere in bird intelligence research. It's a remarkable blending of observation on ecology and behavior, cognitive studies in the lab, and deep-down probing of bird brains themselves aimed at solving the mysteries of the bird mind.

two

THE BIRD WAY

The Avian Brain Revisited

Once when I was cross-country skiing in the Adirondack Mountains, I stopped for lunch in a small clearing. The snow was thick on the ground and the cold was bone chilling. The instant I pulled my peanut butter sandwich from its foil, I caught movement out of the corner of my eye and heard a familiar *ƶeeeee*. A black-capped chickadee (*Poecile atricapillus*), a relative of that cream-skimming tit, had popped up on a branch at the edge of the clearing. Then another, and another. Soon there was a little herd of birds at my feet. When I held a crumb on my finger, one bird fluttered by and plucked it off. Then, moments later, the cheeky little thing perched right on my arm and ate directly from my hand.

You might not tap the chickadee as the brightest bulb in birddom. It's known primarily for its cuteness. Round, fluffy little body in handsome gray coat and smart black cap. Short bill. Oversized head like ET. It has nothing of the slender sleekness of warblers or vireos or the swaggering canny of the crow. It's mostly famous for its vim at the feeder and its astonishing acrobatics. As the ornithologist Edward Howe Forbush once observed, "I have seen a chickadee drop over backward from a branch in pursuit of an insect, catch it, and turning an almost complete somersault in the air, strike right side up again on the leaning trunk of the tree."

But the chickadee is more than just the bird of verve and agility. It's also acrobatic in its aptitudes, curious, intelligent, and opportunistic, with a remarkable memory: "a bird masterpiece beyond all praise," in the words of Forbush. On Louis Lefebvre's IQ scale, the chickadee's family rates right up there with woodpeckers.

Lately, the high, thin whistles and complex gargle calls of chickadees—the *fee-bee*s, *ʒeee*s, *dee-dee-dee*s, and sibilant *stheep*s—have been parsed by scientists and declared one of the most sophisticated and exacting systems of communication of any land animal. Chris Templeton and his colleagues have found that chickadees use their calls like language, complete with syntax that can generate an open-ended number of unique call types. They use some calls to convey their location to another bird or to twitter news of a tasty treat; others, to warn of predators—both the type of beast and the magnitude of its threat. A soft, high-pitched *seet* or sharp *si-si-si* signals a threat on the wing, a shrike or a sharp-shinned hawk. The signature *chickadee-dee-dee* flags a stationary predator, a raptor perched in the treetops or an eastern screech owl looming on a limb above. The number of those skipping-stone *dee*s indicates the predator's size and hence the degree of threat. More *dee*s means a smaller, more dangerous predator. This may seem counterintuitive, but small, agile predators that can maneuver easily are a greater menace than larger, more cumbersome ones. So a pygmy owl may elicit four *dee*s, while a great horned owl may garner only two. These are calls for reinforcements, used to recruit other birds to harass or mob the menace in a group defense calibrated to the magnitude of the threat. So reliable are the chickadee's vocalizations that other species heed their warnings.

Knowing this changes the way I hear those *dee*s as I walk through the woods. Maybe I'm being scrutinized as I pass, sized up and judged dangerous or not.

Or maybe not. Maybe I'm summarily dismissed as a lumbering oaf—big but harmless—my presence causing barely a ripple in conversation.

Chickadees are generally unfazed by people. Bold and inquisitive like bullfinches, they possess a "deep-rooted self-confidence," and will investigate everything inside their home territory, *Homo sapiens* included. They'll hang around the cabins of hunters during hunting season to peck at the fat on the carcasses slung into truck beds. Often, they're the first to visit bird feeders and will even—as I learned—take food straight from the hand. Like bullfinches, they excel at discovering and exploiting new sources of food. Chris Templeton once watched a chickadee feed on nectar from a hanging hummingbird feeder. In winter they'll eat bees, roosting bats, tree sap, and dead fish.

When gallflies were introduced to the American West in the 1970s to help control the spread of invasive spotted knapweed, chickadees seized the new opportunity. Templeton discovered that the birds quickly learned to spot the knapweed seedheads that harbored the highest densities of gallfly larvae—an unusually rich food. Whatever cues they used were subtle, captured on the wing, with little to no time spent hovering over the plants deciding. And yet almost invariably they found the seedheads with the major lode. They would snatch their prizes in flight and carry them back to a tree to pluck out the larvae.

Templeton was astounded. "That chickadees are able to make such successful decisions with so little time spent assessing seedheads is remarkable," he notes. Equally impressive is how quickly the birds learned to exploit a completely novel food source, an exotic insect living on an introduced plant that had been present in its habitat for only a short time.

Chickadees are also possessed of a prodigious memory. They stash seeds and other food in thousands of different hiding places to eat later and can remember where they put a single food item for up to six months.

All of this with a brain roughly twice the size of a garden pea.

NOT LONG AGO I found a chickadee's skull in a patch of scruffy pines near my home. It nested neatly in my palm, chalky white and impossibly light, like the thinnest eggshell. It seemed little more than a bulbous eye case connected to a stitch of beak as sharp as a needle. At the back were twin domes, bubbles of translucent bone, where the brain had been.

A chickadee weighs 11 or 12 grams; its brain, a mere 0.60 or 0.70 gram. How can such a diminutive brain be capable of such complex mental feats?

Clearly there's more to a brain than mere size. But the truth is, birds have long had a bum rap in the size arena. Contrary to the cliché, the brains of many birds are actually considerably larger than expected for their body size. This is the result of an extraordinary process that also gave rise to our own oversized brains—although through a completely separate evolutionary path.

Bird brains range in size from 0.13 gram for a Cuban emerald hummingbird to 46.19 grams for an emperor penguin. Tiny indeed next to the 7,800-gram brain of a sperm whale, but compared with animals of roughly the same size, not so small at all. The brain of a bantam bird weighs about ten times as much as the brain of a similar-sized lizard. Consider a bird's brain relative to its body weight, and it comes out more like a mammal.

Our brains weigh about 1,360 grams, or 3 pounds, for an average body weight of 140 pounds. Wolves and sheep have about the same body weight as we do, but their brains are about one seventh the size of ours. New Caledonian crows are like us, extravagant rule breakers. They possess a brain weighing 7.5 grams in a body that weighs only a little more than half a pound. That's the same size as the brain of a small monkey, like a marmoset or a tamarin, and 50 percent larger than the brain of a bush baby—all animals with roughly the same body size as the crow.

And the brain of a chickadee? It has double the brain size of birds in the same body-weight range, such as a flycatcher or a swallow.

When you look at it this way, many bird species have surprisingly large brains for their body size. They're what scientists call hyperinflated, like our brains.

FOR CENTURIES we thought birds' brains had been reduced in size for good reason: so the marsh hawk could float in wide circles, so the chimney swift could live a life entirely on the wing, so the chickadee could beat a change of course in less than 30 milliseconds.

Brain tissue is heavy and metabolically expensive, the most expensive in the body, second only to the heart. Neurons may be small, but they're costly to make and maintain, consuming about ten times more energy relative to their size than other cells. No wonder nature had trimmed a bird's gray matter, we thought. "Ironically the power of flight, which we regard as the bird's most magnificent attainment, has also been the evolutionary adaptation which has held it well behind the mammals in intelligence," Peter Matthiessen once wrote. Birds solved problems, we assumed, not by relying on their cleverness but by flying away from them.

Flight is indeed an energy hog. In a bird the size of a pigeon, it consumes roughly ten times as much energy as rest. For a small bird such as a finch, short flights that involve a lot of flapping expend almost thirty times as much energy. (By comparison, swimming for a waterbird such as a duck takes about three or four times the energy of rest.) To meet the constraints of flight, nature has in fact considerably lightened a bird's load with a skeleton that blends strength and airiness. Some bones have been fused or eliminated. A light beak made largely of keratin has replaced a heavier, toothy jaw. Other bones, such as wing bones, are pneumatic, almost hollow but reinforced with strutlike trabeculae to keep them from buckling. A bird's bones are dense only where needed—even denser than the bones of their mammal counterparts—in the legs and in the deep solid breastbone that anchors the wings. (So powerful is the downstroke of a bird's wing that it produces force enough to lift twice the animal's body weight into the air.) When biologists examined the genes involved in a bird's skeletal system, they found that birds possess more than twice as many genes for bone remodeling and resorption than mammals do. Most bird bones are hollow and thin walled, yet surprisingly stiff and strong. The paradoxical result sometimes boggles the mind: A frigate bird with a seven-foot wingspan has a skeleton that weighs less than its feathers.

Evolution has found other ways to streamline or totally eliminate a bird's unnecessary body parts. Bladders have been done away with. The liver has dwindled to a mere half gram. A bird's wild knot of a heart is four-chambered and double-barreled like our own, but tiny, with a beat far more rapid (between 500 and 1,000 times a minute for black-capped chickadees; 78 for humans). Its respiratory system is quite extraordinary, proportionately larger than in mammals (one fifth of its body volume, compared with one twentieth in mammals), but much more efficient. Its "flow-through" lung, encased in a rigid trunk, maintains a constant volume (in contrast with mammalian lungs, which expand and contract in a flexible body) and is connected to an intricate web of balloonlike sacs that store air outside the lungs. Unlike most of its reptilian relatives, birds have only one functional ovary, on the left side; the right one was lost over evolutionary time. Only in the breeding season is a bird burdened with heavy sex organs; for most of the year, testes, ovaries, and oviducts are vanishingly small.

The condensed genomes of birds may also be an adaptation to powered flight. Birds have the smallest genomes of any amniote, the group of animals, including reptiles and mammals, that lay their eggs on land. A typical mammal has a genome ranging from 1 billion to 8 billion base pairs, whereas in birds, it hovers at around 1 billion, a result of fewer repeat elements and a large number of so-called deletion events, in which DNA has been expunged over evolutionary time. A more compressed genome may allow a bird to regulate its genes more rapidly to meet the requirements of flight.

ALL OF THIS featherweight frugality is related to a remarkable evolutionary process begun in the dinosaur ancestors of birds.

Thomas Huxley was among the first to see an evolutionary pathway from dinosaurs to modern birds—an observation, by the way, that did nothing to enhance public regard for bird smarts. Huxley, Darwin's bulldog, the "yellow-faced, square-jawed old man, with bright little brown eyes," as his student H. G. Wells called him, had only limited dinosaur fossil finds to draw from, but he saw birdlike traits in them and dinosaur-like traits in the recently discovered 150-million-year-old bird fossil *Archaeopteryx*. Indeed, Huxley wrote, "If the whole hind quarters, from the ilium to the toes, of a half-hatched chicken could be suddenly enlarged, ossified, and fossilized as they are, they would furnish us with the last step of the transition between Birds and Reptiles; for there would be nothing in their characters to prevent us from referring them to the Dinosauria."

Huxley was right, of course. Birds evolved from dinosaurs during the Jurassic period, 150 million to 160 million years ago. In fact, says paleontologist Stephen Brusatte of the University of Edinburgh, "we find that there is no clear distinction between 'dinosaur' and 'bird'. A dinosaur didn't just change into a bird one day; instead, the bird body plan began early and was assembled gradually, piece by piece, over 100 million years of steady evolution."

It's easy to catch the reptilian in birds. You can see it in their beady eyes and quick darting movements; in the pterodactyl-like wings of a rhinoceros hornbill; in a robin holding up his head in frozen alertness to catch a sound, his expressionless face remindful of a lizard's; or in a great blue heron—the slow heavy wing beat, the snaky finesse of its neck, the hoarse squawks, are all a throwback to dinosaur lagoons. But it baffles the imagination to think that the tiny flashlike chickadee could have arisen from the big beasts of vanished ages.

IN A REMOTE CORNER of northeast China is a slash of land that tells the story of this remarkable transition. During the early Cretaceous, volcanic ash coated the region, creating the fossil-laced Jehol beds of Liaoning, Hebei, and inner Mongolia.

When I was at one fossil site near the tiny village of Sihetun in Liaoning Province nearly two decades ago, the layer-cake formations had been excavated only recently by villagers, and everywhere lay fossils of ancient fish, freshwater crustaceans, and mayfly larvae, imprinted on thin, brittle sheets of siltstone. I had come to chronicle a discovery made a year earlier by a farmer and amateur fossil collector probing the layers of the cliff formation. Embedded there was the fossil of a small creature in classic death pose, head thrown back and stiff tail sticking straight up. It looked like a large lizard, about a foot high and two-legged. Running down its back, however, was something extraordinary: a fuzzy mane of simple hairlike filaments.

The creature was a theropod dinosaur named *Sinosauropteryx*, "Chinese dragon feather," a key link between birds and dinosaurs. (The Theropoda, meaning "beast-footed," was a diverse group of bipedal dinosaurs that ranged in size from the monstrously large *Tyrannosaurus rex* and *Deinonychus* on down to foot-high troodontids.) I watched a photographer work with this little theropod fossil for ten hours a day to "raise" the delicate impression of protofeathers locked in stone. It was astonishing to see the dark filamentous streaks emerge from the dinosaur's tail: primitive plumage.

Feathers were one of those signature features deemed the exclusive province of modern birds. The ancient Jehol beds changed all that. In the past two decades, the beds have turned up a flood of fossil dinosaur specimens about 120 million to 130 million years old, with feathers of all types, from rudimentary fuzz and bristles to full-fledged flight feathers. One group of feathered dinosaurs common at the time, the Paraves (which included those *Velociraptors* of *Jurassic Park* fame), were testing out flight modes, gliding, parachuting, jumping between trees; some vaulted into powered flight, and so birds were born.

Dinosaurs gave rise to chickadees and herons in part through a process of relentless shrinking, a kind of Alice in Wonderland phenomenon known as sustained miniaturization. More than 200 million years ago, dinosaurs began rapidly diversifying in body size to fill new ecological niches. However, only the evolutionary line of dinosaurs leading to birds maintained these very high rates of change. Over a period of 50 million years, the theropod ancestors of birds continuously shrank in size, from 163 kilograms to less than 1 kilogram. Nearly everything got small. Being little and light meant that these bird ancestors could explore new food niches and escape predators by climbing trees, gliding, and flying. They developed new adaptations significantly faster than did other dinosaurs. Their small size, evolutionary flexibility, and certain novel adaptations (efficient insulation from highly developed feathers and the ability to fly and forage over long distances) may have helped birds survive the catastrophic events that killed off many of their dinosaurian cousins—and then go on to become one of the most successful groups of land vertebrates on the planet.

Did their brains shrink, too?

Not so much. The dinosaurs that gave rise to birds possessed so-called hyperinflated brains even before the evolution of flight. The visual centers of their brains had already expanded to control the enlarged eyes and superior vision they used to avoid collisions while jumping from tree to tree, as had the regions of the brain used to process sound and coordinate motion. The avian brain evolved to deal with the sophisticated level of neurological and muscular coordination required to explore new niches and escape predators. In other words, bird brains came before birds, just as feathers did.

How does a creature hold on to a big brain while the rest of its body shrinks? Birds managed the trick the same way we did: by keeping a babyish head and face. It's an evolutionary process called paedomorphosis (literally, "child formation"), whereby a creature evolves in such a way as to retain juvenile traits even after it matures.

When an international group of scientists recently compared the skulls of birds, theropods, and crocodilians, they discovered that in most dinosaurs and crocodilians, the skull shape of the animal changed over a lifetime. "During juvenile to adult growth in non-avian dinosaurs, their toothed snouts and faces expanded but their brains increased proportionally significantly less," explains Arkhat Abzhanov of Harvard University, who worked on the study. "Great examples of this are sauropods and stegosaurs with tiny brains relative to their large bodies." In both proto- and modern birds, on the other hand, the skull maintained its youthful shape as the birds matured, leaving plenty of space for enormous eyes and enlarged brains. "When we look at birds," says Abzhanov, "we are looking at juvenile dinosaurs."

As it happens, we humans may have pulled just such a Peter Pan–like move. As adults, we share the big head, flat face, small jaw, and patchy body hair of baby primates. Paedomorphosis may have enabled us to develop bigger brains, just as it did in birds.

NOT ALL BIRDS have brains big for their body size. As with any group of animals, birds have their eggheads and their pinheads. Recall the comparison of birds of similar size, a crow (with a brain of 7 to 10 grams) and a partridge (only 1.9 grams). Or two smaller birds, the great spotted woodpecker (with a brain of 2.7 grams) and the quail (0.73 gram).

Reproductive strategy plays a role in brain size. The 20 percent of bird species that are precocial—born with their eyes open and able to leave the nest within a day or two—have larger brains at birth than altricial birds. The latter are born naked, blind, and helpless and remain in the nest until they're as big as their parents, and only then fully fledge. Precocial birds, such as shorebirds, typically take to life straightaway. Though their brains are relatively large at hatching—allowing them to catch and eat an insect or run short distances when only days old—they don't grow much after birth, so they end up smaller than the brains of altricial birds.

The same is true of brood parasites such as cuckoos, black-headed ducks, and honeyguides, birds that lay their eggs in the nests of others, sparing themselves the costs of rearing their own young. Their chicks, after chucking out their host's offspring (cuckoos) or killing them (honeyguides), also leave the nest early with a brain that's big enough to allow them to fend for themselves but without much later growth.

Why do brood parasites have such small brains? Louis Lefebvre, who has studied brain size in honeyguides, suggests two possibilities. Maybe these birds need to outpace the developmental schedule of their host species, and as a result, they have evolved smaller brains. Or perhaps being a brood parasite relieves the brain of responsibility for all those things associated with raising your own young. "We humans know how much energy it takes to raise a child," says Lefebvre. "If we just dropped our babies into the nests of chimps, we would save ourselves a lot of information processing."

The 80 percent of bird species that are altricial, such as chickadees, tits, crows, ravens, and jays, among others, may be born small brained and helpless, but their brains—like ours—grow a great deal after birth, in part thanks to the nurturing of their parents.

In other words, nest sitters end up with bigger brains than nest quitters.

BRAIN SIZE is also correlated with how long a bird stays in its nest to apprentice with its parents after fledging; the longer the juvenile period, the bigger the brain, perhaps so that a bird can store all it learns. Most intelligent animal species have long childhoods.

One summer, I watched the leisurely upbringing of five great blue heron nestlings in a dead oak tree on a ten-acre pond in Sapsucker Woods, courtesy of a Webcam installed by the Cornell Laboratory of Ornithology. In the past I had caught quick sporadic glimpses of the nest life of robins, bluebirds, and wrens. But this new technology rent the veil, offering an almost embarrassingly intimate and lingering view of the great blue heron's awkward early childhood days.

I've always loved these herons, wide winged and wonderful in their slow sailing. But I never imagined the joy and wonder of seeing them grow up, up close. Like half a million other voyeurs from 166 countries, I became a heron addict.

Our chat room was a close-knit virtual community under the careful supervision of a bird-room "monitor." Classrooms of children tuned in every morning. Someone with an unspecified pain condition tweeted that observing the herons kept her from going crazy.

Together we watched the chicks hatch in late April, then, sleepy and helpless, snuggle under parental down to weather rainstorms and owl attacks, gobble up regurgitated fish and fall into a stupor after eating, peck at everything—sticks, camera, bugs, their parents' bills, one another— all necessary practice for the accurate and powerful spearing of fish. Consideration consternation arose in our virtual community with the hatching of the fifth and last chick, smaller than the others and less aggressive in feeding:

- "#5 getting nothing. Worrying."
- "#5 clacking more, throwing tantrum. Afraid he's not getting enough to eat."
- Then the monitor: "Why do people always want to make up stories that project tragic fates on #5 when he is doing just fine?"

Where there is no drama, people create it. We can't help ourselves.

- "#5 reminds me of the neighbor boy in *Death of a Salesman*. In act 1, he's this nerdy little dweeb, and in act 2, he's a successful attorney arguing cases before the Supreme Court."

AT NIGHT, I watched the herons sleep. Some birds can go without sleep for long periods. Pectoral sandpipers, for instance, forgo sleep for weeks in favor of constant activity under the perpetual light of the Arctic summer. But most species, including herons, seem to share our need for regular sleep. And their sleep, like ours, appears critical to their developing brains.

Birds experience the same cycles of slow-wave sleep and rapid eye movement (REM) sleep that humans do—patterns of brain activity that scientists believe play a crucial role in the growth of big brains, theirs and ours. (That both humans and birds show the same patterns of brain activity during sleep is very likely the upshot of convergent evolution; other vertebrates closely related to birds, such as reptiles, have completely different patterns.) Birds rarely have REM sleep longer than ten seconds, packaged into hundreds of episodes per sleep period, while humans have several bouts of REM sleep per night, each lasting ten minutes to an hour. But for both mammals and birds, REM sleep may be especially important for the early development of the brain. Newborn mammals such as kittens have much more REM sleep than adult cats. Human babies may spend up to half their sleep in the REM stage, whereas for adults, it's about 20 percent. Likewise, studies show that young owlets have more REM sleep than older owlets.

Perhaps those heron chicks did, too.

Like us, birds also have periods of deep, slow-wave sleep in direct proportion to how long they've been awake. Moreover, in both birds and humans, the brain regions used more extensively in waking hours sleep more deeply during subsequent sleep—another similarity born of convergent evolution. A team of international researchers headed by Niels Rattenborg at the Max Planck Institute for Ornithology recently discovered this convergence in a clever study that made use of birds' ability to do something we can't do: modulate their deep sleep by opening one eye, limiting the slow-wave sleep to only one half of the brain while keeping the other half alert, perhaps to navigate while they sleep on the wing, and certainly to watch for predators (an ability that came in handy for the sleeping herons in the predawn darkness of one April morning, when they were attacked by a great horned owl). The team rigged up a little movie theater for several pigeons, blocked one eye in each of them, and showed them David Attenborough's *The Life of Birds*. After staying awake watching the film for eight hours with a single eye, the birds were allowed to sleep. Studies of their brain activity showed deeper slow-wave sleep in the visual processing region of the brain connected to the stimulated eye.

That both humans and birds show this kind of localized brain effect suggests that slow-wave sleep may play a role in maintaining optimal brain functioning, says Rattenborg. "Overall, the parallels between mammalian and avian sleep raise the intriguing possibility that their independent evolution may be related to the function served by this pattern of sleep: the evolution of large, complex brains in both birds and mammals."

I love this idea, that nature dreamed up the same kind of sleep in both humans and birds, fostering the growth of big brains in creatures so far apart on life's tree.

Tuning in each morning to the waking herons was like reading the next chapter in a great coming-of-age novel. In the fledgling phase of May and June, the chicks took clumsy nest-walking laps while mom and dad hustled to feed their rapidly growing bodies, which shot up from 2.5 ounces at hatching to 5 pounds in only seven weeks. Like a baby in a backpack, the young birds took to watchful tracking of everything that moved—airplanes, geese, bees, their parents stalking the pond and gauging their strike angle. Then there was fledging, the first hop-fly, the first plunge over the edge of the nest (which caused a pitch of excitement in the chat room: "Chick #4 looks like kid on high dive getting up nerve to jump." "I'm completely fixated"). Later, the first stabs at plundering the shallows, often unsuccessful, the dogged persistence nonetheless, the return to the nest by nightfall. All under the watchful eye of their parents, who welcomed them back into the nest and offered supplemental frogs and fish.

Contrast this way of life with the precocial plover, which, after erupting from the egg, is up and running—quite literally—almost as soon as its feathers are dry. It's a trade-off: full functionality at birth versus greater brainpower later.

MIGRATION IS ANOTHER TRADE-OFF. Birds that migrate have smaller brains than their sedentary relatives. This makes sense, as a brain that consumes a lot of energy and develops slowly would be too costly for birds that travel a lot. Moreover, according to Daniel Sol of the Centre for Research on Ecology and Forestry Applications in Spain, innate, hardwired behavior may be more useful to migratory species that move between vastly different habitats than learned, innovative behavior. It may not pay to spend a lot of mental resources gathering information in one place that may not be useful in another.

Here's a surprise: Even within a species, brain size may vary—or at least the size of certain brain parts. Vladimir Pravosudov of the University of Nevada and his team compared ten different populations of black-capped chickadees and found that those living in the harsher climates of Alaska, Minnesota, and Maine have a larger hippocampus—the brain region vital for spatial learning and memory—with more neurons, than their counterparts from Iowa or Kansas. The same goes for mountain chickadees, the tough little cousins of the black-capped, who frequent mountains in the West. Mountain chickadees that live in the colder, snowier conditions of higher elevations have bigger hippocampi than their lower-elevation peers. Those from the highest peaks in the Sierra Nevada, for instance, have almost twice the number of hippocampal neurons than their peers living only 650 yards lower down. (They are also better problem solvers.) This makes sense. At higher elevations, where it stays cold longer, birds must store more seeds and remember where they put them. Recovering caches is not so critical in milder climates, where food is available year-round.

Whatever the size, something remarkable happens in the hippocampus of these scatter-hoarding birds on a routine basis. New neurons are born, adding to—or replacing—the old ones. The reason for this neurogenesis remains a mystery. It may allow the brain to recruit new neurons when it's required to learn new information or it may help to prevent a new memory from interfering with an old one. As Pravosudov points out, chickadees "cache food, retrieve caches, and recache previous food caches on a daily basis, especially in winter, so they need to keep track of the old and new caches." The "interference avoidance" idea suggests that birds may need to separate the events by using different neurons for different memories. Pravosudov has shown that chickadees from populations experiencing difficult climatic conditions—and therefore compelled to cache more—have higher rates of neurogenesis.

In any case, this changeover of neurons has forever altered how we think of vertebrate brains, including our own. We are not born with all the brain cells we will ever have, as scientists long believed. In the hippocampus of humans, too, new brain cells are born and others die. We now understand that this ability to change and renew neurons and the connections between them "provides the brain with the potential to modify itself—to learn, at timescales ranging from milliseconds to minutes to weeks," says Pravosudov. In food-caching birds like the chickadee, this plasticity may allow the birds to meet the mental requirements of a demanding world within relatively limited brain space.

THE CONVENTIONAL NOTION that a bigger brain is always better and more powerful in vertebrates such as birds and mammals was finally put to rest with a simple but ingenious new way of measuring brainpower: counting neurons. In 2014, Brazilian neuroscientist Suzana Herculano-Houzel and her colleagues determined the numbers of neurons and other cells in the brains of eleven species of parrots and fourteen species of songbirds. The brains of birds may be small, says Herculano-Houzel, but they "pack surprisingly high numbers of neurons, *really* high, with densities at least akin to what we find in primates. And in corvids and parrots, the numbers are even higher."

Much depends on where the neurons are. Herculano-Houzel has shown that elephant brains have three times the number of neurons found in the human brain (257 billion to our average 86 billion). But 98 percent of them are in the elephant cerebellum, she says, where they may be involved in control of the trunk, a two-hundred-pound appendage with fine sensory and motor capabilities. An elephant's cerebral cortex, on the other hand, which is twice as big as ours, has only a third the number of neurons found in our cerebral cortex. To Herculano-Houzel, this suggests that what determines cognitive abilities is not the number of neurons in the whole brain but in the cerebral cortex—or its equivalent in birds. In a macaw, for instance, Herculano-Houzel and her team found that nearly 80 percent of the brain neurons are contained within the cortexlike part of the brain, while only 20 percent reside in the cerebellum. This is the opposite of the ratio found in most mammals.

In short, finding large numbers of neurons in the cortexlike structures of parrots and songbirds, especially corvids, suggests a "large computational capacity," say the scientists—which in turn may explain the behavioral and cognitive complexity reported for these bird families.

SIZE IS NOT the only reason bird brains have long had a bad rap; anatomy is another. What little brain a bird had was considered primitive, barely more sophisticated than a reptile's. "Birds were viewed as lovely automata capable only of stereotyped activity," says Harvey Karten, a neuroscientist at the University of California, San Diego, who has studied the avian brain for the past half century.

This anatomical dissing began in the late years of the nineteenth century, with the observations of Ludwig Edinger, a German neurobiologist known as the father of comparative anatomy. Edinger believed that evolution was linear and progressive. Like Aristotle, he ranked creatures in a ladderlike *scala naturae*, from lower, less-evolved fish and reptiles to higher, more advanced animals—with humans, of course, crowning the top. Each species going up the steps of the ladder was an elaboration of an older species. In his view, brains evolved in the same step-stool fashion, moving from primitive brain to more complex brain by adding new parts on top of old parts. The more intelligent new brain regions of higher animals were layered over the less intelligent old ones of lower animals like geological strata, progressing in size and complexity from fish and amphibians, with the most primitive brains, to that pinnacle of evolution, the human cerebrum.

The old, or bottom, brain contained neurons arranged in clusters and was the seat of instinctive behaviors, such as feeding, sex, parenting, and motor coordination. The new, or top, brain consisted of six flat layers of cells enveloping the old brain. It was the seat of higher intelligence. In humans it had grown so vast, it had to buckle and fold to fit inside our skulls.

The new layered top brain was the seat of higher thinking. In Edinger's view, birds simply lacked the hardware necessary for complex behavior. Instead of a layered and folded "top" brain, they had essentially smooth "bottom-brain" structures, composed almost entirely of those old reptilian low-life clusters of neurons. Hence they were primarily creatures of instinct—of hardwired, reflexive behavior—and incapable of high-level intellectual feats.

The names Edinger gave to structures reflected his misguided beliefs. He used the prefixes *paleo-* (oldest) and *archi-* (archaic) to label structures in the bird brain and *neo-* (or new) for parts of the mammalian brain. The "old" bird brain was called the paleoencephalon (now the basal ganglia). The "new" mammal's brain was the neoencephalon (now the neocortex). This terminology—which implied that the bird brain was more primitive than a mammal's brain—severely underplayed the mental abilities of birds. Words will do that. We are a naming species, and what we call things influences the way we think about them and the experiments we deem worthy of doing. Calling regions of the bird brain "paleostriatum primitivum" reinforced notions of primordial dimness and stifled interest in studies on bird learning and brainpower.

Thus, the syllogism went like this:

- The neocortex is the special seat of intelligence.
- Birds have no neocortex.
- Therefore, birds have little or no intelligence.

EDINGER'S VIEW HELD for more than a century, into the 1990s. But beginning in the late 1960s, scientists such as Harvey Karten began to look deeper into bird and mammal brains. Karten and his colleagues peered closely at the cells, cell circuits, molecules, and genes of different animal brains and compared them. They examined embryonic development to see which regions of the brain gave rise to others. They traced the wiring of neurons to understand how they linked different brain regions.

What they found turned Edinger's old notions on their heads. Bird brains are not primitive, undeveloped versions of mammalian brains. Birds have been evolving separately from mammals for more than 300 million years, so it's hardly surprising that their brains look quite different. But they do in fact have their own elaborate cortexlike neural system for complex behavior. In ornithological parlance, it's called the dorsal ventricular ridge, or DVR. It arises from the same region of the embryonic brain during development as a mammal's cortex does—the so-called pallium (Latin for "cloak")—and then matures into a dramatically different architectural form.

At the same time, lab experiments began to show evidence of complex behavior in birds: a pigeon's exceptional ability to discriminate between pictures that showed humans and those that did not, for instance— whether the humans were clothed or naked; the African grey parrot's talent for toting up numbers and categorizing objects; and the expertise of certain corvids for remembering the location of other birds' food caches.

BUT DESPITE ALL these breakthroughs, the bias against bird brains held on, in part because of Edinger's ill-conceived labeling of brain regions.

Finally, in 2004 and 2005 came a manifesto rescuing the anatomical reputation of the bird brain. An international group of twenty-nine experts in neuroanatomy, led by two neurobiologists, Erich Jarvis of Duke University and Anton Reiner of the University of Tennessee, issued a series of papers overhauling Edinger's misguided views and antiquated alphabet soup of brain misnomers. (It was not an easy task. One participant described the challenge of seeking consensus among the bird brain experts as an exercise in herding cats.) The members of the Avian Brain

Nomenclature Consortium not only renamed the parts of the bird brain in light of modern understanding; they also related its structures to parallel structures in the mammalian brain so that bird biologists could talk to mammal biologists about what, in fact, were very similar brain regions in their respective subjects.

"About 75 percent of our forebrain is cortex," says Jarvis, "and the same is true for birds, particularly species of songbirds and parrots. They have as much 'cortex,' relatively speaking, as we do. It's just not organized the way ours is." Whereas the nerve cells in a mammal's neocortex are stacked in six distinct layers like plywood, those in the bird's cortex-like structure cluster like cloves in a garlic bulb. But the cells themselves are basically the same, capable of rapid and repetitive firing, and the way they function is equally sophisticated, flexible, and inventive. Moreover, they use the same chemical neurotransmitters to signal between them. And perhaps most important, bird and mammal brains share similar nerve circuits, or pathways between brain regions—which turns out to be vital for complex behavior. It's the connections, the links between brain cells, that matter in the matter of intelligence. And in this regard, bird brains are not so different from our own.

Irene Pepperberg uses a computer analogy. Mammalian brains are like PCs, she says, while bird brains are like Apples. The processing is different, but the output is similar.

The point, says Erich Jarvis, is that there is more than one way to generate complex behaviors: "There's the mammal way. And there's the bird way."

Consider the workings of working memory—one of the cognitive abilities that the New Caledonian crow named 007 revealed in that eight-stage puzzle of sticks, stones, and boxes. Working memory, also called scratchpad memory, is the capacity to remember facts for a short time while addressing a problem. It's what allows us to remember a phone number while we dial it. And it's what allowed 007 to keep in mind his goal while completing the many steps required to achieve it.

Both birds and humans appear to use working memory in a similar way. In our brains, the process that generates it arises in the layered cerebral cortex. But birds have no layered cortex, so how is information in the crow brain stored from moment to moment?

To find out, Andreas Nieder and a team of researchers at the Institute for Neurobiology at the University of Tübingen taught four carrion crows to play a version of pairs, the game of memory that involves holding an image in mind while searching for its match. They showed the crows a random image. The birds then had to remember this image for a second before selecting the match out of four options by tapping the recalled picture with their beaks. Correct answers were rewarded with a mealworm or birdseed pellet. As the crows performed the task, the scientists observed the electrical activity in their brains.

The crows were pros, accomplishing the matching task easily and proficiently. What was happening inside their brains? In what's called the nidopallium caudolaterale region, an area analogous to the prefrontal cortex in a primate, a cluster of up to two hundred cells that were activated when the birds saw the original image remained activated while they searched for a match. This is the same mechanism that allows humans to hold in mind relevant information while we carry out a task.

Clearly working memory can exist without a layered cerebral cortex. In humans and birds, it "differ[s] only with respect to the presence of a language component in humans," says Onur Güntürkün, a neuroscientist at the Ruhr University Bochum in Germany. "The neural processes generating working memory seem to be identical in both."

BIRDS HAVE FINALLY GAINED new respect. They may be relatively small brained, but they are certainly not small minded.

So maybe the question now is not "Are birds smart?" but "*Why* are they smart?" Especially given the constraints of flight on brain size. What evolutionary forces have been at play in shaping bird intelligence?

Theories abound, but two predominate. One claims that ecological problems, especially related to foraging, helped to drive the expansion of bird brains and boosted cognitive abilities: How do I find enough food throughout the year with the challenge of harsh seasons? How do I remember where I hid my seeds? How do I get at hard-to-get foods? In general, animals facing unforgiving or unpredictable environments are thought to have enhanced cognitive abilities, including better problem-solving skills and an openness to exploring new things.

The other theory suggests that social pressures have propelled the evolution of a flexible, intelligent mind: getting along with others, claiming and defending territory, dealing with pilferers and thieves, finding a mate, caring for offspring, sharing responsibilities. (Even the way the wild ibis swaps the leading role in flying formation during migration suggests a kind of adaptive social cognition, an understanding of reciprocity—one good turn deserves another and serves the good of the whole.)

Another idea, first suggested by Darwin, hints that an animal's cognitive abilities are as much a product of sexual selection as natural selection. Do choosy female birds shape the intelligence of their species?

The answers aren't all in, but crows and jaybirds, mockingbirds and finches, pigeons and sparrows, are providing some tantalizing clues.

three

BOFFINS

Technical Wizardry

A bird named Blue has a problem. Next to him on a table in his aviary is a plastic tube with a piece of meat tucked inside, just out of reach of his beak. Like 007, Blue is a New Caledonian crow, a bird known for its masterful tool craftsmanship and keen problem-solving skills.

Blue scopes out the situation, hopping around the tube, peering inside, moving his head with click-stop precision. He flutters down to the floor of the aviary and pecks around at various random objects scattered there—leaves, tiny twigs, a stray piece of plastic or two—but apparently doesn't find what he's looking for. He flies to a scrubby spray of branches clustered in a pot on the table and perches, cocking his head to the right and then the left, surveying his options. He picks a twig and snaps it off the bough. Then he methodically snips off all the side twiglets. Now he has a nice long, straight stick. The right tool for the job. He jabs the stick into the tube and spears the meat, then polishes it off.

Watching Blue neatly fashion a perfect little tool from scrubby branches is a wonder. In the wild, these crows make elaborate tools from sticks, leaf edges, and other materials, which they use to winkle grubs and insects from burrows in fallen wood, from behind bark or leaves, and from the base of leaves, crevices, holes, and cavities of all kinds. The crows travel with their tools, suggesting they value them; they know a good tool when they see one and keep it for reuse.

There's something almost outlandish about this behavior. Birds making a tool so good they want to reuse it? Plenty of animals use tools. But few make such elaborate ones. In fact, as far as we know, only four groups of animals on the planet craft their own complex tools: humans, chimps, orangutans, and New Caledonian crows. And even fewer make tools they keep and reuse.

THIS SCENE WITH BLUE is a little window on a big idea. Birds are smart because they have to solve problems in their environment—specifically, problems of how to get food from hard-to-get places. It's called the technical intelligence hypothesis. Ecological challenges have provided an evolutionary stimulus to intelligence in birds.

In British slang, *boffin* is a term for a kind of technological geek, someone with extensive skill in a specialized field. The New Caledonian crow fits the bill. Its use of tools is unmatched in the bird world. And its ability to craft its own tools is on par with such primate smarties as chimpanzees and orangutans.

Why does it matter? What's the big deal about tools?

We once considered the ability to make and use tools to be a sign of high intelligence or complex cognition, uniquely human, like language or consciousness. The use of tools, we thought, required some distinctively human understanding, including causal reasoning, a grasp of cause and effect. It was one thing that singled us out as special and played a pivotal role in our evolution and development as a species. Benjamin Franklin called us *Homo faber*, "man the toolmaker." A list of the tools we have invented "is a useful proxy for the entire history of our species," according to Alex Taylor and Russell Gray of the University of Auckland: "Stone axes, fire, clothes, pottery, the wheel, paper, concrete, gunpowder, the printing press, the automobile, the nuclear bomb, the internet: The manufacture of these tools created revolutions in the societies they were invented in, as each of them either redefined how humans interacted with either the environment, or with each other."

The notion of tool use as uniquely human went by the wayside when Jane Goodall discovered that the chimpanzees of Gombe National Park also use tools. So do orangutans, macaques, and elephants, it turns out, and even insects. A female digger wasp will hold a pebble in her mandibles and use it to hammer shut the soil and pebbles sealing her burrow. Weaver ants harness their own larvae as tools in building and repairing their sturdy nests. The worker ants pick up the larvae, which secrete silk, and shuttle them back and forth so the silk cements together the leaves in their nests. Still, tool use is exceedingly rare in the animal world, documented in less than 1 percent of species.

For a long time, primates were considered the primo tool users. But in the past decade or so, New Caledonian crows have emerged as contenders for that title. This is no small achievement. Especially when you look at the catalog of, say, orangutan tools, which range from toothpicks and teeth cleaners to autoerotic tools and missiles aimed at predators, from leaf napkins and moss sponges to leafy branch fans and scoops, chisels, hooks, nail cleaners, and bee covers—branches or leaves used as a hat to protect against stinging bees. Or a chimp's clever constructions: a "rake" made by combining as many as three sticks or bamboo poles to reach a reward, or a kind of plate assembled from leaves and then reassembled to form a drinking cup.

Even in this shrewd company, New Caledonian crows stand out. While they may not make and use the variety of tools that chimps and orangutans do, they craft their tools with precision from a range of materials. They make them the proper length and diameter for any particular task. They modify them to solve new problems. They innovate. They use tools in sequence, as 007 did in the video of that eight-stage puzzle, using a short tool to get a longer one that can be used to get food. And perhaps most impressively, they make and use hook tools—the only species other than humans to do so.

THE FIRST TIME I saw a New Caledonian crow use a tool in the wild was along a road that climbs steeply between Focalo and Farino in southern New Caledonia. At an overlook along the road, the government had recently erected an elaborate wooden guardrail. The spot pulls tourists from the highway to gawk at the stunning views of forested mountains and the blue waters of Moindou Bay. But that April morning, the spot was more popular with winged visitors.

Alex Taylor has taken me there in the hopes of catching the crows nut cracking during their morning feeding. The birds have a fairly strict daily rhythm not unlike our own eight-hour day: They're active from dawn through late morning, depending on the heat, then they take their version of a siesta until early afternoon, when they get active again until dusk.

"Right now, they're in businesslike food-gathering mode," explains Taylor. "It's a small window in the day when they're willing to put themselves in danger."

Sure enough, four or five families of New Caledonian crows are rustling around in the shrubs below the road, flitting from branch to branch, quietly *wak-wak*-ing on the ground. Someone has dumped a load of garbage over the edge of the road, and the birds are sorting through the offal.

New Caledonian crows, like rats and humans, are euryphagous, partial to a variety of plant and animal food. They will happily consume insects and their larvae, snails, lizards, carrion, fruit, nuts, and human leftovers littering a site. The pickings here are easy, and it seems unlikely that the crows will bother with anything as demanding as nut cracking. The nuts they feed on belong to the candlenut tree—the same tree that harbors the juicy beetle larvae the crows extract with their tools—and they're not easy to open. But suddenly we hear a sharp crack on the pavement behind us. We turn around and notice several crows in the trees above the road.

One perches on a forked branch overhanging the pavement, releases a nut—which breaks with a *thwack*—and then swoops down to collect the nutmeat from the broken shell.

Not only do New Caledonian crows crack nuts in this way, but the more gourmand among them have also been known to split escargots by this method, tossing the rare endemic snail *Placostylus fibratus* onto the rocky beds of dry creeks in the rainforest to get at their tasty insides.

Many birds open nuts, shellfish, and eggs in similar fashion. The vampire finch of the Galápagos Islands breaks apart the large eggs of boobies by bracing its beak on the ground and booting the egg with both feet to crack it against rocks or roll it over a cliff. The black-breasted buzzard of Australia unloads stones on emu nests, and the Egyptian vulture releases them onto ostrich eggs. Carrion crows use passing cars to crush especially tough nuts, such as walnuts, that won't break by simply falling on pavement. The now-famous video of these crows in a city in Japan shows one stationed above a pedestrian crossing. When the light turns red, it positions its nut on the crossing, then flies back to the perch and waits while the light changes and traffic passes; when the light turns red again, it flutters down to safely collect the cracked nut. If no car smashed the nut, the bird repositions it.

Strictly speaking, dropping food on hard surfaces isn't tool use. But the New Caledonian crows here have added a twist. Down the road from us, a crow lands on that newly built wooden barrier. He drops his nut into a big round hole in the wood that houses a large metal bolt. He lodges the nut there, then uses the bolt as an anvil to hold it firmly while he pries open the cracked nut with his beak. Ingenious.

OTHER BIRDS USE TOOLS they find lying around. A wanderer in the pages of ornithological journals and Robert Shumaker's fascinating compendium on *Animal Tool Behavior* discovers delightful and surprising accounts of birds using found objects as tools—to contain water or scratch their backs or wipe themselves down or lure prey: for instance, a white stork bringing water to its chicks in a clump of damp moss and then wringing it out to fill their beaks; African grey parrots bailing water from their dish with a tobacco pipe or a bottle cap; an American crow ferrying water in a Frisbee to dampen its dried mash, and one securing a plastic Slinky toy onto its perch and using the free end to scratch its head; a Gila woodpecker fashioning a wooden scoop out of tree bark to carry honey home to its young; a blue jay using its own body as a napkin to rid ants of their noxious formic acid spray, making them fit for eating.

Some birds use objects as weapons. One American crow in Stillwater, Oklahoma, lobbed three pinecones at a scientist's head as he climbed up to its nest. A pair of ravens in Oregon defending their nestlings from two intruding researchers used similar tactics but harder weaponry. "A rock the size of a golf ball fell past my face and landed next to my feet," wrote one of the scientists. The researchers assumed that a raven perched on a cliff above the nest had accidentally kicked loose the stone. But then they saw a raven with a rock in its beak. With a quick flip of its head, the raven tossed the rock down toward its target. Then six more, one after another. A rock that struck a researcher on the leg had marks indicating that the raven had pulled it out of the ground, where it had been partially buried.

Several kinds of birds use objects as lures to draw fish. Green-backed herons are expert bait fishers, known to entice their prey with bread, popcorn, seeds, flowers, live insects, spiders, feathers, even pellets of fish food. Dung is the decoy of choice for the burrowing owl. The owls scatter clumps of animal feces near the mouth of their nest chambers and wait motionless like muggers for unsuspecting dung beetles to scuttle toward their trap.

Nuthatches hold bark flakes or scales in their bills to lever the bark from trees, exposing the bugs beneath. One chestnut-backed chickadee was seen using a thorn to pry seeds from a suet feeder. Other notable bird uses of sticks, twigs, and branches: as drumsticks by black palm cockatoos, which regularly use them in the wild to thrum a hollow tree trunk for territorial display or to direct a female's attention to a possible breeding hole; as back scratchers (as well as head, neck, and throat scratchers) by yellow-crested cockatoos and African grey parrots; as a club by a bald eagle seen bludgeoning a turtle with a stick held in its bill; and, perhaps most unusual, as a kind of bayonet in a scuffle over seed between a crow and a jay.

This last example is the first documented case of a bird using an object as a weapon against another bird, so it's worth pausing to explain. Early one April morning not long ago, ornithologist Russell Balda was watching an American crow leisurely feeding at a platform in Flagstaff, Arizona, that was stocked daily with a variety of seeds for local birds. Steller's jays visited the site often to take advantage of the easy fare, flying off with the seeds to cache them nearby. One jay, apparently unhappy with the crow's unhurried pace of dining that morning, tried to dislodge the bigger bird by scolding and dive-bombing it, but to no avail. The jay then flew into a nearby tree and vigorously worked with its bill to break off a twig from a dead branch. It succeeded and, taking the blunt end in its beak with the sharp end pointing outward, flew back down to the platform. Brandishing the twig like a lance or spear, it lunged at the crow, missing its body by an inch. When the crow lunged back, the jay dropped the twig. The crow picked it up, pointed end outward, and stabbed back at the jay. The jay flew off, with the crow in hot pursuit, twig still in bill.

THESE ARE MOSTLY EXAMPLES of sporadic tool use. Among the handful of birds other than New Caledonian crows that make a regular habit of using tools are the woodpecker finches (*Cactospiza pallida*) of the Galápagos Islands.

One of the many finches Darwin found in the Galápagos, each with a beak shape optimized for the most abundant food source on the islands it occupies, the small, buff-breasted woodpecker finch has a powerful picklike bill that it uses to chip away at bark and old wood to uncover grubs and beetles. Splints of wood fly off during the work, and the birds use these to probe tree holes or crevices beyond the reach of their beaks.

They also use twigs, leaf petioles, and cactus spines to pull arthropods from otherwise inaccessible nooks and crannies. Sabine Tebbich, a behavioral biologist at the University of Vienna who has studied these birds for more than fifteen years, has found that only those finches living in dry, unpredictable habitats, where food is scarce and hard to access, use tools—and will spend half their foraging hours doing so. Woodpecker finches that live in more humid areas, where food is more abundant and easy to obtain, on the other hand, rarely use tools.

In the first experimental study of how birds may acquire their use of tools, Tebbich found that woodpecker finches are born with the ability and don't need an adult tutor to refine their skills, but they do grow more skillful over time through trial-and-error learning.

One bird Tebbich acquired for her study offered a close look at its gradual mastery of tool skills. Tebbich found Whish in a domed nest of mosses and grasses tucked into the branches of a giant Scalesia tree on Santa Cruz Island. The chick was only a few days old and crippled from attack by fly larvae. Over the next several months, a small army of scientists at the Charles Darwin Research Station cared for him; two of them documented his progress in a charming account.

At first the finch showed little interest in objects. But when he was almost two months old, he began to play with flower stems and small twigs, twiddling them in his beak and holding them at right angles to his bill. Soon he was investigating everything around him with great curiosity, tweaking buttons, nibbling pencils, yanking hair through the small ventilation holes in a slouch hat, prying apart toes with his beak and tools, inspecting ears and earrings. By three months, he was an accomplished tool user and had broadened his toolkit, probing cracks with twigs, a feather, fragments of water-worn glass, wood slivers, shell pieces, and the hind leg of a large green grasshopper. He also inserted a twig between a sock and a boot.

"It seemed to Whish that any potential crack was worth trying to open," wrote the scientists. "Even a person's face was not sacrosanct. He would fly to the face and clutch hold on the nose arch. He would then hang upside down and peer into the nostrils. If the face possessed a beard he would sometimes land on the hair, as if on a mossy trunk. From this vantage point, he thrust his bill between the lips and forced them apart. If the mouth opened, he then examined the teeth with the tip of the beak."

Recently Tebbich and her colleagues observed two woodpecker finches in the wild, an adult and a juvenile, doing something new: The birds found a novel kind of tool and modified it for better effect. The adult bird tweaked several barbed twigs from blackberry bushes and removed the leaves and side twigs. It oriented the tool so the barbs were facing in the right direction to effectively drag its prey—arthropods—from under the bark of a Scalesia tree. The juvenile bird watched the adult using the tool and then used it himself in the same way.

One gets the feeling that other birds out there may be better boffins than we think; we just haven't caught them in the act. Take Goffin's cockatoos (*Cacatua goffini*), small white cockatoos with a "bishop's hat" crest, known to be curious and playful and, in captivity, exceedingly adept at picking locks. No one has observed these birds using tools in their natural tropical dry forest habitat on the Tanimbar archipelago of Indonesia. But Alice Auersperg and her team at the University of Vienna watched a captive cockatoo named Figaro spontaneously tear long splinters from the wooden beams of his cage with his beak; he then used a splinter to rake in a nut just out of reach. In later experiments, Figaro fashioned a new tool for every nut placed out of reach, "successfully, reliably, and repeatedly" making and modifying stick-type tools using different materials and techniques to accomplish the task.

STILL, AS FAR AS WE KNOW, in terms of artful toolmaking and use in the wild, no bird matches the New Caledonian crow.

A few years ago, Christian Rutz of the University of St Andrews and his team used motion-triggered video cameras to get detailed views of the crows using tools at seven sites in the wild. Over a period of about four months, they recorded more than 300 site visits and 150 instances of crows using tools to extract larvae from wood. The dexterity of the crows is astonishing. A crow fishing for larvae looks a lot like the termite fishing Jane Goodall observed in the chimpanzees of Gombe. The crows repeatedly poke the larvae with their tool until the creature chomps down on the tip with its powerful mandibles. By carefully wiggling the tool, flicking gently right and left, twisting it ever so slightly, the bird brings a grub to the surface and carefully withdraws the tool without losing its prey. It may sound easy, but it's not, even for us humans with our nimble digits. Rutz and his colleagues tried their own hands at the task and found that it requires "remarkable levels of sensorimotor control" and is "surprisingly difficult to master."

When it comes to the nuts and bolts of tool crafting, only chimps and orangutans match or exceed the New Caledonian crow's sophistication. And not even these hotshot primates can make hook tools. As if that were not enough, the crows make not one but *two* kinds of hook tools—one from live twigs and the other from the barbed edges of the leaves of pandanus trees, or screw pines.

Show-offs.

The twig hook tool involves snipping one side off a forked twig and breaking off the remaining side just below the base of the fork, then removing all the side twiglets. From the short stump that remains, the crow sculpts a little hook, sharpening the point until it's perfectly suited to fishing out small prey.

The pandanus tools are made from the spiny strap-shaped leaves that crown the pandanus tree. They come in three designs: wide, narrow, and stepped. The stepped version is the most sophisticated, says Alex Taylor. It's wide and sturdy at the top, easy to grip, and steps down to a thin and flexible probing end. It takes many complex moves conducted in a very precise manner to complete the tool—snipping at one spot and tearing along that edge, then snipping at another spot and tearing from there, several times in a row. The final version looks a lot like a miniature saw but is used as a probe to wheedle out grasshoppers, crickets, cockroaches, slugs, spiders, and other invertebrates from otherwise inaccessible nooks and crannies.

One remarkable feature of these implements: Unlike tools made by other animals—such as the brush-tip tools of chimpanzees, created in sequential steps—the complete shape and design of a pandanus stepped tool is determined before it's made. The bird crafts the whole thing while it's still on the leaf. It works as a tool only after the crow makes a final cut to separate it from the leaf. This suggests to some scientists that the bird may be working from some form of mental template.

And another cool thing: Once the tool is removed from the leaf, an exact negative impression of its shape remains on the leaf, a "counterpart." In an island-wide survey, Gavin Hunt and Russell Gray of the University of Auckland studied the shapes of more than five thousand counterparts at dozens of sites across New Caledonia. They found that the styles of toolmaking varied from place to place, and those styles seem to have persisted for decades. In some parts of the island, the crows make wide tools primarily. In others, more narrow tools. The stepped-tool design is the one that's the most widespread over the island. On the island of Mare, just adjacent to New Caledonia, says Hunt, the crows make only wide tools. In other words, it seems there may be local styles or traditions of toolmaking that are passed down over generations.

Faithful transmission of local tool designs: If it's true, that fairly well defines the term *culture*.

Moreover, in Hunt's view, there's evidence that the crows have made incremental improvements in their tool designs over time—which would make them the only nonprimate species known thus far to demonstrate "cumulative technological change." At most sites in New Caledonia, the crows make only the stepped-tool design that is the most complex of the three types of pandanus leaf tools. "I think it's highly unlikely that a naïve crow without any experience of pandanus tools could have invented a multi-stepped tool without first making a simpler tool," says Hunt. And yet there's no sign of the more basic designs on pandanus leaves at these sites. "The birds don't appear to make earlier, simpler designs," says Hunt; "they just seem to go straight to making the most complex design—just as humans go straight to making the latest model and don't recapitulate all the technological stages that enabled them to get to the current design." Circumstantial evidence, to be sure, but "we often accept parsimonious explanations in the absence of absolute proof," says Hunt. In his view, the evidence points to cumulative improvement in pandanus tool technology.

Christian Rutz argues that there is not yet sufficient evidence to justify these claims; more study is needed. However, the crows do seem to understand how their hooked stick tools work, which hints at how cumulative improvement may have come about. In a suite of experiments on wild-caught New Caledonian crows, Rutz and his colleague James J. H. St Clair found that the birds paid close attention to which end of a tool is hooked and oriented it correctly. This recognition, they write, "has implications for the timescales over which tools may be profitably curated." That is, the birds can reuse a tool even if they don't remember which way they put it down and can also use tools discarded by other birds—which, say the scientists, "is potentially a key mechanism for the social learning and diffusion of tool-related information in crow populations." Moreover, the pair argues, the crows' ability to distinguish between the functional features of tools and modify them—make them a little better—may contribute to the evolution of tool complexity.

WHY IS IT THAT out of some 117 species of corvids, the New Caledonian crow became such a wizard with gizmos? What were the forces nudging the crow toward this remarkable faculty? Other crows are smart. Other crows live in tropical locations. Is there something special about this place? This bird?

New Caledonia is by all measures a wondrous spot. A remote sprig of land 220 miles long between New Zealand and Papua, from the air the island looks like it was born of the same fiery forces as other Pacific islands, Hawaii or Bali or nearby Vanuatu: tall green mountains, white beaches, blue lagoons. But unlike most islands dotting these warm seas, New Caledonia is not young and volcanic. It's the geologic offspring of the ancient supercontinent Gondwanaland, the northernmost edge of an almost completely submerged continent, Zealandia, which rifted away from Australia 66 million years ago. It was underwater until its final emersion 37 million years ago.

The island is one of the quietest places I've ever been to. It has roughly the land area of New Jersey, but less than 3 percent of the population, so in many places it feels almost uninhabited. The Kanak, the island's indigenous people, make up more than two fifths of the population, while the Caldoches, the Europeans—mostly French—make up about a third; the rest are a mix of peoples from neighboring islands. The empty roads are frequented by those big swamphens known as pukekos, with bright red bills and violet chests. Tall, slender Cook pines pillar the sky, named for the renowned explorer Captain James Cook. When Cook approached the island—one of the first Europeans to do so—in 1774, he and his crew saw "a vast cluster . . . of elevated objects" and bet on whether they were trees or stone pillars. The pines belong to a tree family often called living fossils because they look like the ancestral evergreen trees that forested the planet in dinosaur days. Down the center of the island runs a spine of mountains, the eastern slopes patchy green with primeval rainforest. In the gloom below the canopy lives that ghostly kagu, a bird that may be a relict species dating from the days of Gondwanaland.

The primeval rainforest that once blanketed New Caledonia has been reduced to pockets. But the island remains a diversity hotspot, with insect species thought to number upward of 20,000, including more than 70 native species of butterflies and 300-plus species of moths. There are some 3,200 species of plants on the island, three quarters of which are endemic, found nowhere else. For this reason, New Caledonia is often considered its own distinct floristic subkingdom.

It is also an ark of colossal creatures. The giant gecko, the "devil in the trees," for instance, which measures fourteen inches, and skinks that reach a hefty twenty-three inches. One mammoth air-breathing land snail, *Placostylus fibratus*, grows to a full five inches. The goliath imperial pigeon, known locally as Notou, is the world's largest arboreal pigeon, weighing in at more than 2.2 pounds—roughly twice the heft of a common rock dove. Gone the way of extinction is the ground-bound swamphen *Porphyrio kukwiedei*, a bird the size of a turkey, and the huge flightless *Sylviornis neocaledoniae*, five and a half feet long and 66 pounds.

Strange things happen on islands. Gigantism is not uncommon. Nor is dwarfism or gaudy experiment or anomalies of every kind. On the island of Borneo, I caught sight of a male Asian paradise flycatcher, a bird no bigger than a robin, but dangling a pair of weirdly elongated central tail feathers, opalescent streamers a foot long that rippled through the vivid green of the rainforest like the tail of a kite.

Islands are castles of experiment surrounded by moats. Competition is less fierce and predators less abundant than on continents, so evolutionary experimentation is not so quickly or ruthlessly punished. That includes behavioral experimentation, like tooling around with tools. (Perhaps it's not surprising that the only other birds on the planet to regularly potter with tools are the woodpecker finches of the Galápagos.)

According to Christian Rutz and his colleagues, the crows probably arrived on the island of New Caledonia sometime after its emersion 37 million years ago. Some fossilized skulls and bones of the crow have been excavated in Mé Auré Cave in the Moindou region of the island. But these remains are only a few thousand years old, so they're not much help in understanding the bird's deeper evolutionary history.

The corvid family split into different lineages tens of millions of years ago, but the ancestral line of New Caledonian crows is probably not that old. It's likely that ancestors of these crows flew large distances over open water to reach New Caledonia, Rutz suggests, probably from Southeast Asia or Australasia. Modern New Caledonian crows are poor fliers; they typically make only short perch-to-perch flights, and when they have to go longer distances, their flight is slow and labored. But Rutz suspects they descended from birds that were either strong fliers or lucky colonists. And most likely, it was after these crows colonized the island that they evolved their extraordinary ability to make and use tools—sometime in the past several million years.

FOR ANIMALS CLEVER ENOUGH to extract it, New Caledonia offers a concealed cache of rich, juicy prey: grubs of the long-horned beetle and other invertebrates that burrow deep into wood. The grubs are heavy in protein and energy-rich lipids. According to Rutz, a crow can satisfy its entire daily energy requirements with only a few larvae. There's not much competition for these natural energy packets. No woodpeckers or monkeys or apes or aye-ayes or striped possums or other so-called extractive specialists that can pull food from holes.

Nor are there slews of enemies to threaten the crows from earth or sky. The island does have some aerial predators—the whistling kite, the peregrine falcon, the white-bellied goshawk—but these are not generally considered threats to the crows. New Caledonia has no snakes to speak of (apart from the burrowing blind snake, which lives only on the smaller islands adjacent to the main island) and no native predatory mammals. The island's only native mammals are nine species of bats that play a major role in dispersing the seeds of many rainforest tree species. When Cook arrived on the island—naming it New Caledonia after his beloved Scotland—he brought with him two dogs as a present for the Kanak people. Bad idea. Now feral dogs abound, along with other introduced species, such as cats and rats. The dogs have decimated the kagu population, but they pose little danger to the crows.

One consequence of such modest threat from competitors and predators is that the crows are free from the burden of vigilance—in other words, they have the time and ease of mind to tinker with sticks and barbed leaves, to poke and probe, to bite and tear, and then probe again, without looking up. Freedom from threats may also have allowed for the evolution of a more leisurely childhood, in which young crows under the watch of their parents could dabble safely in toolmaking, refining their skills over a long period of time without starving in the process.

BABY CROWS don't jump out of the nest able to make perfect tools. Some evidence suggests that they're genetically predisposed to use them, as woodpecker finches are. One experiment showed that young crows in captivity learn to make and use basic stick tools on their own, without exposure to adults. But when it comes to making more complex tools, young birds clearly benefit from adult tutoring or modeling.

They make full-fledged pandanus tools, for instance, only after spending time with an adult. The learning process tends to be a school of hard knocks, softened by the presence of a doting parent or parents. As a PhD student studying with Russell Gray and Gavin Hunt at the University of Auckland, Jenny Holzhaider spent two years in the rainforests of New Caledonia observing how young crows learn to make and use pandanus tools in the wild. Watching the videotapes she and Gray made of a bird named Yellow-Yellow (for the double yellow bands he wears) is a little like watching a toddler learning how to eat with a spoon without spilling his food. It's a slow process, full of mishaps and lost opportunities.

In his lecture on the evolution of cognition, Gray describes the young bird's progress. At the start, Yellow-Yellow has no clue what he's doing. By the age of two or three months, he's paying close attention to the actions of his mother, Pandora. He sees her fish with her tool. Then he borrows the tool and tries to wedge it down a hole sideways. He seems to get what the tool is for but not how to use it. By using her tools and following her around, he's learning what sorts of plants and sticks make good fishing implements and roughly what tools are good for.

Once he starts trying to make his own tools, he doesn't mimic his mother's movements. Rather, he emulates the tools she makes, attempting his own approximate copies. This may help to explain the existence of "regional" styles of toolmaking. By observing and using tools made by their parents, juvenile crows "may form a kind of mental template of the locally produced tool design and use this as a basis for their own tool manufacture," explains Gray. "We know that in birdsong, there's a form of template matching where through trial-and-error learning, the juveniles match the adult song. Maybe the same kind of neural circuits are being coopted for template matching in toolmaking."

The rest of the process appears to be mostly a matter of experimentation. Over the next months, the young bird tries his hand (or bill) at ripping up pandanus leaves, willy-nilly. The bits and pieces he tears off seem random—but at least he's getting the hang of the tearing technique.

At five months, he's capable of making something that looks like a tool. But he often uses the wrong part of the pandanus leaf, the part with no barbs, so the tool is useless. He flips it around and tries using it that way, but no-go. A few months later, he's got the "manufacturing" sequence down and is executing all the right moves—cutting the pandanus leaves in the proper area, ripping away pieces step-by-step, all very carefully. But because he started in the wrong spot, the tool is upside down, with barbs facing in the wrong direction.

Half the tools Yellow-Yellow makes won't bring him any food. It's almost a year and a half before he's practiced at making adultlike pandanus tools that allow him to feed himself effectively. That's a long stint of schooling. It works only because his parents support his education by letting him tag along and use their tools, and when he fails at feeding himself, they pop a fat grub or two in his beak to tide him over. The island does its part by allowing him to spend long hours of his young life honing his skills, moving gradually from bumbling apprentice to amateur tinkerer to expert toolmaker without interruption from, say, death.

In this respect, New Caledonian crows may offer clues to understanding our own human strategies of life. We humans stand out in our primate tribe for the extended period of juvenile dependence we enjoy and our learning-intensive survival strategies. According to the Auckland team, the link between a high level of technological skill in foraging and a long juvenile period of provisioning by parents in both humans and New

Caledonian crows suggests that the two traits may be causally related. It's called the early learning hypothesis. Perhaps possessing learning-intensive tool skills plays a role in lengthening the juvenile period. In this way, New Caledonian crows may provide a good model for investigating the evolutionary effect of tool use on life history, not just for birds but for people.

A HIDDEN TROVE of rich food, scarce competition, and a paucity of predators may have created conditions favoring tool crafting, but as Christian Rutz points out, these factors alone are not sufficient to produce it. Lots of crows in the Pacific region with similar lifestyles and access to pandanus leaves don't make tools. In northeast Australia dwells the Torresian crow, a cousin of the New Caledonian crow. It lives among the larvae of the Australian longicorn beetle and has no competition for this superrich food source, but it hasn't figured out how to tap it with tools. Nor is the white-billed crow of the Solomon Islands, perhaps the New Caledonian crow's closest relative, a tool user.

Is there something special in the physical or mental makeup of the New Caledonian crow? Something in its body or brain that distinguishes it from its corvid colleagues?

I FIRST MET the bird early one morning when I stepped out of my lodge in La Foa, in central New Caledonia.

There it is, a few feet in front of me, in the low branches of a scrabbly tree. In a way, I'm delighted to see that it doesn't differ so much from the American crows that haunt my neighborhood. Ebony of beak, legs, and feathers. Upper feathers glossy, iridescent purple, deep blue, or green, depending on the light. Body about the size of . . . well . . . of a smallish crow, though a bit more compact than our American crows and beefier than your average jay or jackdaw.

The bird cocks its head at me. Its eyes are large and prominent, dark brown, beady, intelligent. They are also positioned more toward the front of the head, capable of rotating around and orienting forward during tool use to create a kind of extraordinary binocular "overlap" greater than that of any other bird. This wide, binocular visual field allows the crow to position its bill accurately when probing with a tool.

New research by Alex Kacelnik of Oxford University and his colleagues suggests another visual twist. In crows, as in humans, one eye is more dominant than the other. Crows hold their tools on one side of the bill or the other so they can see the tip of the tool and their target with their preferred eye. As Kacelnik puts it, "If you were holding a brush in your mouth and one of your eyes [was] better than the other at [gauging] brush length, you would hold the brush so that its tip fell in view of the better eye. This is what crows do."

The bird's bill, for its part, is straight, conical, and businesslike, free of fancy hooks or curves that mark other corvid beaks—the better to firmly grip a tool and bring its tip into range of that powerful binocular field.

A beak is the one appendage birds have for investigating the edible world. Ordinarily, its shape seriously limits what a bird can eat. Hawks and eagles have hooked beaks for ripping into rabbits. Herons have tonglike bills for snapping up slippery fish. Woodpeckers have sharp picklike beaks for chipping away at wood. Some crows have hooks; others have pincers; still others have spears.

On its own, the bill of the New Caledonian crow can do only so much. But this crow has figured out how to expand its reach through the miracle of tools.

It's not clear which came first, toolmaking or the unusual physical adaptations so neatly adapted to its requirements. Did the crows' bill shape and specialized vision predispose them to tool crafting and use? Or did their tool-use behavior in response to unusual ecological opportunity—those delicious hidden grubs—gradually shape their visual system and bills? This is the sort of mysterious causal relationship biologists both love and hate.

In any case, say scientists, these two features—specialized visual system and straight, conical bill—allow for a level of tool control impossible for other corvids and are akin to the features that enabled our own skillful tool handling, including binocular vision and flexible wrists and opposable thumbs, which allow us to grip and pinch with precision.

As Gavin Hunt points out, several other aspects of the New Caledonian crow's toolmaking lifestyle also resemble our own. There's the extraordinarily long extended juvenile period with parental care, which scaffolds the learning of toolmaking and tool use. Moreover, says Hunt, "in both humans and crows, tool use is both genetically inherited and flexible, and thus widespread if not universal. The occurrence of tool use appears to be patchy across both species. Thus the transmission process, even if it involves less social learning in New Caledonian crows than in humans, produces a very similar result."

THE CROW stares back at me, intense, quizzical, as if asking what is so surprising. I wonder if the brain inside that black skullcap is any different from the brain of other corvids. Research suggests that there may be small differences. One study shows that the New Caledonian crow's brain is bigger, at least compared with the brains of carrion crows and European magpies and jays. (However, as we know, overall brain size can be a rudimentary measure.) There is some ballooning in areas of the forebrain thought to be involved in fine motor control and in associative learning. This may improve the crow's dexterity and boost its ability to pay attention to what it's doing—a big advantage in any mental challenge. Moreover, as Russell Gray points out, New Caledonian crow brains have slightly higher numbers of glial cells, which in humans are thought to be involved in the mechanism of learning and memory known as synaptic plasticity. In sum, the brains of these crows may possess "no miraculous extra new structure," observes Gray, "just small, incremental tweaks."

But are these crows capable of high-level thinking? Can they understand physical principles such as cause and effect? Do they reason and plan and make leaps of insight?

For the past decade or so, the University of Auckland team and their colleagues have been probing the crow's mind, poking around its nooks and crannies to see what kind of special understanding, if any, these birds may possess. They're less interested in trumpeting the crow's overall intelligence than in exploring what they call the "signature" cognitive mechanisms at play when the bird solves problems. These may be the building blocks or foundations of such sophisticated human cognitive abilities as insight and reasoning, imagining and planning. They include skills such as the ability to notice the consequences of one's own actions, to grasp cause and effect, and to assess the physical characteristics of materials.

"When these birds are problem-solving, they may be using forms of cognition intermediate between simple learning and human thought," explains Taylor. The signatures of cognition evident in the crows' behavior might represent the in-between steps along the way to our own complex cognitive abilities such as imagining scenarios or reasoning about cause and effect. "That's why we're really interested in these crows as model species," says Taylor. "Pinpointing the cognitive mechanisms they use can offer insights into the evolution of human thinking and of intelligence in general."

Consider what 007 did in that video of the eight-stage metatool puzzle. It looked like the clever crow solved the problem with insight. He seemed to study the whole problem—"There's food in that box I can't reach with my beak"—and then, by playing out a complex mental scenario in his head, to solve the problem in a flash of understanding, planning his sequence of moves and executing them one at a time, keeping his ultimate goal always in mind.

According to Russell Gray, who conducted the original metatool experiments with Taylor, what 007 did was probably less sensational than this—though still intriguing. The bird did indeed scope out the problem, says Gray. But he probably wasn't using his imagination or building scenarios in the way that we do, or solving the problem in a flash of insight. Instead, he was acting on objects that were physically present and familiar to him. He knew how they worked. He paid close attention to how his tools interacted with the other objects. Drawing on his past experience with the objects, he followed an appropriate sequence of actions that led to his goal. If he was using mental scenario building at all, suggests Gray, it was of a highly limited kind—dependent on context and experience.

The actions of 007 could be either more sophisticated than this or even simpler, notes Alex Taylor, "a type of moment-to-moment decision-making, with no mental simulation at all," he says. "We just don't know. These are competing hypotheses we need to test."

THE AVIARY WHERE the University of Auckland conducts its mind-probing experiments sits in a scrubby field behind a small research station in Focalo. In the wet season, a creek meanders through the property, subject to extreme flooding during storms; its bed is dry now, shaded by shambly melaleuca trees and the occasional pandanus. Except for the low, husky *waaa*-ing of the seven crows currently occupying the netted enclosures, the scene is quiet. Horses wander through the field, once in a while setting off raucous crow alarm calls when they come too close.

A stream of well-studied crows have flowed through this aviary, among them 007 and now Blue, named for the blue band he wears on his left leg. The Auckland team keeps the birds in the aviary for a few months before releasing them back into the wild (007, for instance, was released back into his home forest on New Caledonia's Mount Koghi). The colored bands help them to keep track of which bird is which, and offer a stand-in for a name until imagination kicks in with something more inventive. After christening more than 150 birds (Icarus, Maya, Lazlo, Luigi, Gypsy, Colin, Caspar, Lucy, Ruby, Joker, Brat, to name a few), Alex Taylor says he has exhausted his supply and asks for suggestions. So Blue's daughters, Red and Green, are now named after my daughters, Zoë and Nell.

The scientists capture the crows using a "whoosh" net and try to collect them in family groups. In places with high-density populations of the birds (say, twenty crows per square mile), this is fairly straightforward.

But in many spots on the island, especially in higher-elevation forests, the birds are more sparsely distributed (more like two or three crows per square mile) and may be particularly difficult to catch. Taylor's colleague Gavin Hunt recently had a devil of a time collecting birds in the Mount Panié area. It was the official Kanak hunting season of the Notou. New Caledonian crows occasionally end up in the line of fire aimed at the pigeons, so the crows are more skittish than usual during this season. Hunt came away empty-handed. Even without the disruption of gunshots, the process takes patience.

Once the captured crows are brought to the aviary, they quickly adjust to their new digs. And who wouldn't? Taylor and his colleague Elsa Loissel feed them fresh ripe tomatoes, cubes of beef, papaya, coconut, eggs. ("People have the mistaken impression that science is all about thinking and experimenting," Loissel quips, "when a lot of time is actually spent chopping tomatoes or cutting beef into tiny cubes.") Before long, the birds settle in and come whizzing down to the table for work. "The trick is keeping them amused," says Taylor, "keeping the rhythm going of making their tasks just hard enough for them to stay interested and engaged."

"What we really want to understand is how these crows think," says Taylor. How do they solve complex problems? Through insight or reasoning or something more mundane?

Think back to the string-pulling task that was part of 007's eight-step challenge. The crow's remarkable ability to spontaneously pull up a stick attached to a string hanging from a perch has been viewed by some scientists as proof of insight. The bird creates a mental simulation of the problem (imagining the effect that pulling up the string will have on the position of the food) and then instantly executes a plan to solve it.

To see whether this is so, Taylor and his colleagues set up a variation of the experiment using a string with meat attached for reward. In their version, the crows couldn't see the meat moving toward them when they pulled the string. This stymied the birds. Without the visual reinforcement of the meat moving closer and closer, cuing them to keep up the activity, only one crow out of eleven spontaneously pulled the string a sufficient number of times to get the meat. Their performance stooped to canine cluelessness. (It should be noted that humans goof at this, too: Scientists tested this string connectivity task with fifty undergraduates, says Taylor, and nine of them failed.) When the birds were given a mirror in which to watch their progress, they once again excelled at solving the problem. If this had been an example of insight, of a sudden, instantaneous understanding of cause and effect—pull on string and meat will move closer—the birds wouldn't have needed visual feedback to continuously direct their actions.

Whether New Caledonian crows have leaps of insight remains to be determined, but these experiments suggest that these birds do have an extraordinary ability to notice the consequences of their own actions, says Taylor, and to pay attention to the way objects interact. These are mighty useful mental tools when it comes to making and using material tools.

THE AUCKLAND TEAM IS also attempting to figure out whether the crows understand basic physical principles. A "crow-appropriate paradigm" for this, as Taylor puts it, is an experimental version of the old Aesop's fable "The Crow and the Pitcher."

In that fable, a thirsty crow comes across a half-filled jug of water. Unable to reach the water to drink, the crow drops pebble after pebble into the pitcher until the water level rises enough for him to drink.

As it turns out, this is not just a folktale. New Caledonian crows will do exactly that—drop stones into a water-filled tube to raise the water level. And as Sarah Jelbert discovered while working with the Auckland team, if given a choice between heavy objects and light ones, solid and hollow ones, the crows will spontaneously pick objects that will sink over those that will float. They know how to pick their materials and will select the right option 90 percent of the time. This suggests that the crows understand water displacement, a fairly sophisticated physical concept, on par with the comprehension of a child five to seven years old. It also suggests that they're able to grasp the basic physical properties of objects and make inferences about them.

Lately, Taylor and Gray and their colleagues have been trying to worm out whether the birds understand the relationship between cause and effect, especially the effect of forces they can't see. This is called causal reasoning, and it's one of our most powerful mental abilities. Causal reasoning is at the root of our understanding that objects in the world behave in predictable ways and that mechanisms or forces we can't see may be responsible for events. "We're constantly making inferences about things we can't see," says Gray. If we're standing inside and a Frisbee flies through our window, we understand that someone must have thrown it. The human ability to reason about causal agents develops very early in life. An infant only seven to ten months old shows surprise if a beanbag is thrown from behind a screen and then the screen is lifted to reveal a toy block rather than an expected human causal agent such as a hand. As Gray points out, this ability underlies our grasp of thunder and head colds, magnets and tides, gravity and gods. It also helps us to understand the behavior of people around us and allows us to make and use tools and adapt them to novel situations. It's another one of those potent capacities once thought unique to humans.

Can crows make similar inferences about forces they can't see, so-called hidden causal agents? The idea for an experiment to test this notion was actually suggested to Alex Taylor by a crow.

SCIENTISTS WHO STUDY bird behavior live chancier lives than many of their colleagues, at risk of being foiled by the creatures they're studying—or, with any luck, instructed by them. Birds can undo the cleverest devices, taking the stuffing out of a scientist about as fast as anything you can imagine. But there are other moments—if one is paying attention—that may offer a rich reward. In this case, the surprising behavior of a crow named Laura offered inspiration for Taylor.

It was during an early stage in the Aesop's experimental trial. Taylor baited a floating cork with food, then dropped the cork into a tube of water. He always performed this task with his back turned to the crows. The typical scenario that followed was this: Once the birds solved the puzzle, raising the water level to get at the treat, they immediately flew up to a perch at the back of the cage, ripped the meat off the cork to eat it, and then dropped the cork. To rebait the cork, Taylor had to retrieve it from the rear of the cage. "Which is fine for one trial," he says. "But after a hundred trials, you're fairly sick of that." The task of retrieval is made all the more difficult because the aviary is set up to suit the crows, "nice wide table, plenty of perches," says Taylor. "So it becomes kind of a jungle, impossible for a human to move through. You end up crawling on your hands and knees a lot."

Laura did things differently. Like the other birds, she took off with the baited cork, but once she had eaten the meat, she would fly back to the table and leave the cork there, very close to where Taylor was standing. "I was like, 'Oh, thank you so much. That is so awesome!'" Not only was he grateful that he didn't have to crawl under the table to retrieve the cork but he could rebait it quickly, speeding up the pace of the experiment.

This made Taylor think. Maybe Laura had figured out his role as the causal agent responsible for the food offering—even though she had never actually seen him bait the cork. "I thought, maybe she understands that if she gets the cork back to me, she gets the food more quickly. She's really good at this task; I'm the limiting factor. So if she can speed me up, she speeds up her treats."

Laura's behavior made Taylor wonder if New Caledonian crows might have a more sophisticated grasp of causal reasoning than we imagined. Do they understand that humans can act as causal agents even when their actions are hidden? Can they reason about unobservable causal mechanisms?

To find out, he and his colleagues devised an inventive experiment. The idea was to figure out whether the crows could infer that the movement of a stick poking in and out of a hide was caused by a person they had seen entering the hide. In an open aviary, the team set up a hide behind a tarp. On a table next to the hide sat a small box holding food that could be extracted by a crow with a simple tool. To get at the food, the crows had to turn their backs to the tarp hide. In the tarp was a hole. When a stick was thrust through the hole from behind the hide, it poked directly into the space where the crow's head would be when it was probing for food in the box, posing a clear danger.

For the experiment, eight crows watched two different scenarios in which the stick poked through the hole, explains Taylor. The first scenario was the hidden-causal-agent situation: A human entered the hide, the stick moved in and out of the hole several times, and then the human left the hide. In the second scenario, no human entered or exited the hide, but the stick still poked in and out.

After observing each scenario, the crows were given the chance to probe for food in the box. Their behavior suggested that they were able to connect the dots and infer that the hidden human was causing the stick's movement. When the birds watched the stick move and then saw the human leave the hide, they appeared comfortable flying down to the table and turning their backs to the hide so they could probe for food. However, when they watched the stick move with no apparent cause, they behaved more skittishly, flying to the table, but inspecting the hide nervously and sometimes abandoning their probing, as if they suspected that whatever unknown force had moved the stick might make it move again. (This is not unlike the surprise an infant shows when a beanbag appears to have been thrown without a human hand.) The difference in the crows' behavior, say the scientists, suggests that the birds may be capable of a fairly sophisticated form of causal reasoning.

In another experiment, this one on "causal intervention," the crows didn't fare so well. Causal intervention is a step beyond causal understanding. It involves seeing something transpire in the world and then acting to create the same effect. Say, for example, you've never shaken a fruit tree to release the fruit. But one day, you see the wind blowing a branch, causing the fruit to fall. And from that observation, you infer that if you shake the branch, you can act like the wind and make the fruit fall.

A gadget called a blicket box offers just this sort of challenge. It's a little box that plays music when you put an object on top of it. Give a two-year-old child a quick demonstration of this, then give her the box and the object and ask, "Can you make it go?" and she will have no trouble re-creating the effect. But crows fail at the task. "All they have to do is pick up the object and pop it on the box," says Taylor. "It sounds so simple to human minds. It's like, duh, how hard could that be, right? But the crows don't get it."

Taylor finds these crow failures as intriguing as their successes. If you're interested in the evolution of cognitive mechanisms, it's equally interesting to see where birds fail, he explains. "We're trying to understand which parts of causal understanding may have evolved together and which did not," he says. "I'm not out to cheerlead these crows. I just want to know how their minds work. If they end up being 'stupid' in some areas and smart in other areas; if they can't do some things but can do others—that's just as interesting. The cool thing about them is their wild behavior and their tool use. That's what defines them."

TAYLOR ADMITS TO interest in another line of inquiry. Less academic, perhaps, but no less intriguing: What do New Caledonian crows do for fun?

"My impression is that they're kind of workaholics," he says. "They're very focused on getting food, but once they've got the food, they just chill out, sit and do a bit of preening, a bit of flying, a bit of calling. But they're not constantly playing with new things, like keas do. I find that fascinating because everyone always says curiosity and play are linked with intelligence."

Do birds play? Do they do things just for fun?

Nathan Emery, senior lecturer in animal intelligence at Queen Mary, University of London, and Nicola Clayton of the University of Cambridge suggest that larger-brained, altricial species of birds (like many mammals) do play—although it "seems to be relatively uncommon in birds," they write, "seen in only 1% of the approximately 10,000 species and largely restricted to species with an extended developmental period, such as crows and parrots."

Play is not necessarily only about preparing a bird for later life, say Emery and Clayton. It may reduce stress, aid social bonding, or just induce pleasure. "Birds, like us, may also play because it is fun," they explain; "it produces a pleasurable experience—releasing endogenous opioids." That is, play can be its own consummatory act, self-rewarding.

According to zoologist Millicent Ficken, it's only clever birds that are capable of complex play activities. And through play, they make discoveries and experiment with the relationship between their own actions and the external world. In other words, play both requires intelligence and nurtures it.

Members of the parrot family tend to be an irrepressibly playful lot. When my parents bought a pet parakeet for our family many decades ago, they also purchased a menagerie of toys to outfit his cage: ladders, mirrors, bells, all made of brightly colored cheap plastic, as well as several strangely shaped food treats. It was standard procedure at the time. Gre-Gre, as we called him, played with every new device until it broke from overuse. These days, pet stores sell whole lines of special parrot toys. African grey parrots prefer playthings like toilet paper rolls, junk mail, Popsicle sticks, paper cups, and plastic pen tops, anything made of paper, cardboard, wood, and rawhide that they can shred, chew, or otherwise destroy. Sometimes they get so lost in their fun, they fall off their perches.

According to expert testimony, the kings of fowl play are keas. These crow-sized parrots live in the Southern Alps of New Zealand. They're nicknamed "mountain monkeys" because of their cheeky nature and primatelike intelligence. On the derivation of their Latin name, *Nestor notabilis*, one book offers this account: "Nestor was a legendary Greek hero known for long life and wisdom, and the name is often used to mean a wise counselor, a leader." Then the killjoy comment: Linnaeus bestowed the name on this parrot family, probably "without thought of any special significance."

Maybe. Maybe not.

Two scientists, Judy Diamond and Alan Bond, who have studied the kea for many years, consider it possibly the smartest, most waggish bird in the world.

"Play in keas is less a set of ritualized behaviors than an attitude to the world at large," they write. When it comes to playing with stuff, keas far outshine their corvid cousins. They are "bold, curious, and ingeniously destructive," says Diamond, considered (depending on who you ask) either playful comics—"clowns of the mountains"—or destructive hoodlums that go around in juvenile gangs trashing things, deconstructing windshield wipers and the vinyl trim on cars, as well as campers' tents and backpacks, rain gutters, and outdoor furniture. The kea's playfulness with objects may help them develop a "toolkit" of behaviors for dealing with novel situations or unexpected foraging problems.

Keas also love to horseplay. An invitation to another bird occurs in the form of a cock of the head and a kind of stiff-legged sidling up to a potential play partner. The partners parry and duel with their beaks, ducking, thrusting, ducking again. They tussle, lock bills, bite, push with their feet, roll on their backs while squealing and waving their feet, and stand on each other's stomachs. There are no winners or losers. (Everyone gets a trophy.)

Sometimes keas play the imp or practical jokester. According to Diamond and Bond, the birds have been known to steal television antennae from houses and deflate automobile tires. One kea was observed rolling up a doormat and pushing it down a flight of steps. A few years ago, the New Zealand *Sunday Morning Herald* reported that a kea stole eleven hundred dollars from an unsuspecting Scottish tourist. At a rest area near the highest pass over the Southern Alps, Peter Leach had rolled down the windows of his camper van to snap pictures of the views and an odd green bird on the ground near his vehicle. Before he knew it, the bird had fluttered into his van. It snatched a small cloth bag from his dashboard and zipped off with it. "It took all the money I had," Leach said with chagrin. "The birds are now lining their nests with £50 notes."

Keas may be the titans of tomfoolery, but corvids frolic, too. Ravens play toss with themselves, throwing twigs up in the air and catching them. Two young white-necked ravens were once observed playing "king of the castle," one standing on a mound brandishing a lump of dung while the other charged up and tried to seize the object.

One clear sunny February morning in the central mountains of Hokkaido, Japan, naturalist Mark Brazil noticed two ravens on a steep slope of fresh powdery snow. One of the ravens lay on its breast and slid down the slope; its partner rolled, legs in the air, wings flicking. "The pair continued this 'sledging' and rolling downhill for more than ten metres before flying back upslope," wrote Brazil, then repeated the hotdogging. Crows, too, have been known to slide down slopes, apparently for fun. Carrion crows were caught on camera in Japan skidding down a children's slide. Not long ago, a video from Russia of a crow snowboarding down a roof with a jar lid went viral.

Alice Auersperg and an international team of scientists recently took a close look at how various species of crows and parrots play with toy objects to see if the nature of their play might shed light on the cognitive nature of the players—as well as the relationship between play and tool use. Playing with objects often precedes using them as tools in primates as well as in birds. A survey of seventy-four primate species found that only tool users, like capuchin monkeys and great apes, combine objects when they play. Children start smashing objects together when they're eight months old. By ten months, they can insert toys into cavities or stack rings on a pole. But it's only after the age of two or so that they begin using objects as tools to achieve a desired goal.

The researchers gave nine species of parrots and three species of crows the same sets of wooden toddler toys of various shapes (sticks, rings, cubes, and balls) and colors (red, yellow, and blue). They also gave them an "activity plate," a kind of playground of tubes and holes for inserting the objects or stacking the rings.

Most of the birds interacted with the toys, but a few were champion players. New Caledonian crows, cockatoos, and keas were most apt to combine two free toys and to use toys on the "playground." The most complex object play, say the researchers, occurred in species with the highest performance in technical innovation and tool use—Goffin's cockatoos and New Caledonian crows. The Goffin's favored yellow toys (which may have something to do with the fact that these birds have yellow stripes beneath their wings, an area often used for social display); the New Caledonian crows, for no apparent reason, preferred balls above all other objects, but they did like to poke sticks into the cavities on the playground. Only Goffin's and young New Caledonian crows combined three free objects, and only parrots stacked rings onto tubes and poles, the Goffin's more than any other, neatly coordinating their beak with one foot to accomplish the task. These Indonesian birds are known for their outstanding problem-solving skills and for using tools creatively in captivity.

"Our studies show a link between object play and functional behavior in these large-brained birds," says Auersperg. "But the direct role that play behavior plays in their problem-solving abilities remains unclear. It might serve as general motor skill practice or learning about object affordances"—the relationship between the object and the bird or the object and its environment, which "affords" the bird an opportunity to perform an action. "Or," she says, "it could just be a side product of their exploration mode."

Interesting to note: All the birds seemed happy to share while they played. No bird hogged more than one activity plate or more than two or three toys at once. "There were no overt cases of aggression and monopolization of objects was not pronounced," say the researchers.

Taylor observes that New Caledonian crows in his aviaries don't seem to play for play's sake. "They like having things in their beaks, bits and bobs," he says. "If you put tools in the cage, they'll spend a lot of time caching the stick, picking it up, probing into things with it. But it's hard to call that play, because in the wild, that's how they make their living."

Recently, Taylor was interested in finding out whether New Caledonian crows might be motivated by a little spontaneous, self-rewarding fun instead of food. His lure: a pair of tiny skateboards to see if these crows, like their Japanese and Russian cousins, might enjoy sliding. Unfortunately, the experiment didn't pan out. "They really didn't like sliding," says Taylor, "so we rather gave up on this."

ONE SERIOUS QUESTION the Auckland team and other scientists would like to tease from the crow's mind is this: Which came first, tool use or these impressive cognitive abilities? Did using tools make these birds smarter? Or were they supersmart to begin with, and their cognitive abilities provided a kind of "platform" or mental toolkit for figuring out how to use tools?

It's possible that life on the island fostered intelligence in these birds, as it may have for the woodpecker finches of the Galápagos. A relatively unpredictable environment may have created evolutionary pressure to evolve sophisticated cognitive abilities for coping with its challenges. These adaptations, in turn, may have provided a foundation for the evolution of tool use.

On the other hand, tool use itself may have driven the evolution of sophisticated cognitive abilities. Maybe the crows chanced on the use of a stick tool to extract food. This exposed them to new sorts of mental challenges that stimulated their ability to solve physical problems. The tool users had a selective advantage because they could get at these fabulously rich grubs. (So rich a food source are grubs that the kaka, a New Zealand parrot, will spend more than eighty minutes extracting a single grub with its long bill.) Once the technique spread, natural selection may have favored the evolution of traits such as extreme binocular vision, which improved its efficiency.

According to Alex Taylor, this chicken-and-egg question is a holy grail for New Caledonian crow researchers: "If it's the case that sophisticated tools affect intelligence, then the populations that have a history of making more sophisticated tools would be smarter. And that would provide evidence for the technical intelligence hypothesis."

Of course, as Gavin Hunt points out, the birds had to have some mental sophistication to put two and two together to come up with the idea of using a tool. "Still, I'm not sure whether New Caledonian crows would have been cleverer than other crow species initially," Hunt says. "But once they initiated tool use, that drove the enhancement of their cognitive skills to the level we see today, which is quite impressive."

So perhaps tool use is not unlike play: It both requires intelligence and nurtures it.

THE BIRD DUBBED 007 came from the forests of Mount Koghi, where the crows make sophisticated hook tools. Was he exceptional in any way? "In terms of his boldness and willingness to persevere, yes," says Taylor. "He was a young bird from a family of three—all pretty keen, on the ball." One researcher who worked with 007 would simply point at him, and the bird took this as his signal to come down for a work session. Sometimes Taylor would find 007 queuing up at the door to his aviary, keen on getting started with his work. "I'd have to say to him, 'So sorry you have to wait; I'm testing the dumb birds down the corridor!'"

But Taylor finds the variation between individual crows less interesting than the differences between populations of crows from different parts of the island, how they vary in their tool use and cognitive abilities.

Next up for the University of Auckland investigators: joining an ambitious international effort to explore the genetic basis of the New Caledonian crow's intelligence as a whole and the differences between populations. One approach involves comparing the genomes of New Caledonian crows and other closely related species. The plan is to identify genes that may have been selected for in the New Caledonian crow lineage but not in closely related species—and then to see how those might be linked to differences in cognitive abilities.

Another approach, now under way in the Auckland aviary, is looking within the New Caledonian crow population for variations in cognitive abilities and genes. A bird like 007, for instance, who comes from the Mount Koghi population of crows that make hook tools, may carry variants of genes that differ from the genes of Blue, who comes from the basic stick-tool-making population of La Foa, in central New Caledonia. Do crows from different parts of the island, with different types of tool manufacture, differ in their cognitive abilities? And do these differences correlate with genetic variations?

MY LAST DAY in New Caledonia, I drive up a narrow switchback road to the top of Mount Koghi, 007's birthplace. The primeval rainforest that cloaks Koghi's slopes is known as the home of Goliaths, Leach's giant gecko and the towering Koghi Kauri, a tree of massive girth, as much as eight feet across, that pierces the canopy at sixty or seventy feet.

According to Taylor, 007 probably has his own family by now. I'm hoping to catch a glimpse of the Koghi crows, but the day is waning. I'm used to the slow, ruddy burnishing of dusk. Here at the equator, the day clangs shut in sudden finality, especially in the dim pitch of the rainforest. Suddenly, the woods are spooky.

Every forest has its own character, its own whispered rumors and smells. The primeval mountain forests of New Caledonia hold echoes of early plants, early birds. In the moist, shaded understory grows the evergreen shrub *Amborella*, the nearest relative to the world's first flowering plants. Huge primeval tree ferns from the Cyatheaceae family, like those that grew in the Permian period 275 million years ago, reach heights of sixty-five feet, with fronds up to ten feet long, among the largest leaves in the plant kingdom. In Kanak languages, the name of the tree fern means "the beginning of the country of men." Creation stories tell of the first human ancestor climbing out of a hollow tree fern trunk.

Time seems to move in a different dimension here. Hurry drifts into the radiant greens. The mind is quieted by wonder.

Walking along, peering up into the thick canopy, my binoculars trained on the lower branches, I stumble on a root and launch into a massive spiderweb. That's when I notice the ungodly abundance of spiders in this forest, orb weavers, I think, that spin elaborate radial webs, which glimmer gold in spears of sunlight. Here in the dim wattage, I can barely see them, but it seems that every gap between the trees is latticed with their webbing, and in the center of each web lurks a sizable spider, still and vigilant. What flashes through my mind is that *Far Side* cartoon showing two spiders perched in a giant web as a fat little boy walks toward it. One spider says to the other, "If we pull this off, we'll eat like kings."

I pick my way forward more gingerly, moving deeper into the deepening green.

Then, up in a tree just off to my right, I hear the quiet *waaaa, waaaa* begging call used by young New Caledonian crows with their parents. All I can make out is a stirring of leaves. Who knows, maybe 007 is up there, feeding his young with grubs he has gathered with a hook tool. Does the DNA he has passed on to his offspring explain why his kind, of all birds on the planet, makes such elaborate tools? Are his genes as a hook-tool maker different from Blue's?

The New Caledonian crow dossier is still packed with unanswered questions. Which came first, the crow's remarkable tool use or its exceptional intelligence? Toolmaking, or the beak shape and vision so neatly adapted to its requirements? DNA for problem solving, or tricky environmental challenges that shape genes?

These are the kinds of mysterious biological questions that I find cheering—untidy, unresolved, still in process. As darkness falls, it's pleasing to contemplate the mystery. Somehow, time in its stew has mixed island and bird, and slowly, incrementally, over the long pull of evolution, come up with this mind-blowing maker of tools.

Talk about genius.

four

TWITTER

Social Savvy

We "rub and polish our brains by contact with those of others."
—MICHEL DE MONTAIGNE

Many bird species are highly social. They breed in colonies, bathe in groups, roost in congregations, forage in flocks. They eavesdrop. They argue. They cheat. They deceive and manipulate. They kidnap. They divorce. They display a strong sense of fairness. They give gifts. They play keep-away and tug-of-war with twigs, strands of Spanish moss, bits of gauze. They pilfer from their neighbors. They warn their young away from strangers. They tease. They share. They cultivate social networks. They vie for status. They kiss to console one another. They teach their young. They blackmail their parents. They summon witnesses to the death of a peer. They may even grieve.

Not long ago, this kind of social savoir faire was presumed far beyond a bird's reach. The idea, for instance, that birds could think about what other birds might be thinking was considered preposterous. Lately, the view has shifted, with science suggesting that some bird species have social lives nearly as complex as our own, which require some very sophisticated mental skills indeed.

The world's thousands of bird species display a dazzling array of social organizations. Some, such as the belted kingfisher and the scissor-tailed flycatcher (also known as the Texas bird of paradise), are solitary and ferociously territorial, hanging out only in mated pairs. Others are born for group company: rooks, for instance, supremely social Old World members of the crow family that nest in crowded rookeries from the United Kingdom to Japan; or king eiders, large ducks of Arctic coastal waters that love to mingle and gather in prodigious collective flocks of up to ten thousand.

Great tits (*Parus major*), the colorful little yellow-breasted birds widespread across Eurasia, have an intriguing social organization that gives new meaning to the old adage "birds of a feather." Researchers from Oxford University recently constructed a kind of Facebook for the tits, an "association matrix" revealing the pattern of associations among individuals in a population of a thousand great tits in Wytham Woods, a stretch of well-studied old woodlands to the west of Oxford. The study revealed just who affiliates with whom and which birds regularly forage at the same place together. The tits, it turns out, have a complicated social network in which birds gather in loose foraging flocks based on their personalities.

Even chickens form complex social relationships. Within a few days of socializing, chickens establish a stable social group with a clear hierarchy. In fact, we owe the expression "pecking order" to studies of the social relations among chickens by the Norwegian zoologist Thorleif Schjelderup-Ebbe, who found that pecking orders are ladderlike, with the top rung conferring great privilege in the form of food and safety, and the bottom rung fraught with vulnerability and risk.

HAS ALL OF THIS living cheek by jowl with mates, family, friends, and peers made birds smart? Do they owe their quick, flexible minds not just to the tricky physical challenges of their environment but to the sticky social ones, the trials and tribulations of getting along? It's called the social intelligence hypothesis, and among scientists, it has lately won a considerable following.

The idea that a demanding social life might drive the evolution of brainpower was developed by Nicholas Humphrey, a psychologist at the London School of Economics, in 1976.

Humphrey was pondering the monkeys housed in groups of eight or nine in his laboratory. The monkeys lived in austere wire mesh cages, and he worried about whether the impoverished environment would affect the cognitive functioning of the young ones. There were no objects, no toys, no environmental stimuli of any kind, and there was no need to avoid predators or search for food (the monkeys were fed regularly). So it seemed to Humphrey that there were no problems to solve. In the face of this, he was puzzled by the monkeys' sharp intellect and their ability to perform impressive cognitive feats despite living day to day in a barren, stultifying environment. All they had was one another.

"And then one day I looked again," Humphrey writes, "and saw a half-weaned infant pestering its mother, two adolescents engaged in a mock battle, an old male grooming a female whilst another female tried to sidle up to him, and I suddenly saw the scene with new eyes: forget about the absence of objects, these monkeys had each other to manipulate and to explore. There could be no risk of their dying an intellectual death when the social environment provided such obvious opportunity for participating in a running dialectical debate."

The rich social milieu "came close to resembling a simian School of Athens," wrote Humphrey, and required unique cognitive skills and social calculations. The monkeys had to gauge the consequences of their own behavior within the group. They had to size up one another. They had to try to guess the likely behavior of their peers, track others' social relationships—dominance, rank, and competitive ability—and weigh advantage and loss in their interactions. All of these calculations were "ephemeral, ambiguous and liable to change," requiring constant reevaluation. It was a game of social plot and counterplot that promoted intellectual faculties of the highest order, argued Humphrey. To interact effectively, social animals had to become "natural psychologists."

———

NOW SCIENTISTS BELIEVE that many bird species are not so different. Those that live in social groups have to sort out social contacts, smooth ruffled feathers, and avoid squabbles. They have to monitor the behavior of others to make decisions about whether to cooperate or compete, whom to communicate with, whom to learn from. They have to recognize numerous individuals, keep track of them, recall what this or that confederate did the last time—and predict what he or she will do now. Because many species of birds share the same kinds of social challenges that may have fueled intelligence in primates, their brains, like ours, may be "designed" to manage relationships.

A range of bird species show impressive social smarts. Magpies recognize their own image in a mirror, a form of self-awareness that we once believed was restricted to humans and a handful of other sophisticated social mammals. When experimenters placed a red dot on the throat of six magpies, two of them tried to scratch off the dot on their own bodies with their legs, rather than reacting to the image in the mirror.

African grey parrots are remarkable collaborators. In the wild, these birds roost in flocks of thousands, forage in groups of thirty or so, and form lifelong bonds with a mate. They're rarely alone—unless they're in captivity. In the lab, they'll pair up to solve physical puzzles, pulling a string together to open a food box. They also understand the benefits of reciprocity and sharing and will opt for a food reward that will be shared with a human rather than one enjoyed solo, as long as they know their human friend will also reciprocate.

Reciprocity in the form of gift giving is another kind of social behavior unusual in nonhumans but fairly common among certain birds, including crows. Two decades ago when a family friend first reported receiving gifts from the crows she regularly fed—a marble, a little wooden bead, a bottle cap, colored berries, all left on her doorstep—I was skeptical. But in recent years, tales have rolled in from all over the country of crows offering up gifts of jewelry, hardware, shards of glass, a Santa figurine, a foam dart from a toy gun, a Donald Duck Pez dispenser, even a candy heart with "love" printed on it, delivered just after Valentine's Day. In 2015, a story surfaced in Seattle of an eight-year-old girl, Gabi Mann, who started feeding crows on her way to and from the bus stop when she was only four. Later she began offering the crows peanuts on a tray in her yard as part of a daily ritual, and from time to time, after the peanuts had been consumed, trinkets showed up on the tray: an earring, bolts and screws, hinges, buttons, a tiny white plastic tube, a rotting crab claw, a small scrap of metal printed with the word "best," and Gabi's favorite, an opalescent white heart. The less "icky" objects Gabi has collected in plastic bags labeled with the dates they were received.

"Leaving gifts suggests that crows understand the benefit of reciprocating past acts that have benefited them and also that they anticipate future reward," write biologist John Marzluff and his coauthor Tony Angell in their book *Gifts of the Crow*. "It is a planned activity; the crow has to plan to bring the gift and plan to leave the gift."

Crows and ravens will balk at doing work for less reward than a peer is getting. This sensitivity to inequity had previously been thought to exist only in primates and dogs and is considered a crucial cognitive tool in the evolution of human cooperation.

Corvids and cockatoos can delay gratification if they think a reward is worth waiting for—a form of emotional intelligence involving self-control, persistence, and the ability to motivate oneself. Young children who can stave off eating one marshmallow now in favor of two later have nothing over these winged marvels of willpower. Alice Auersperg and her team at the University of Vienna found that Goffin's cockatoos offered a pecan would wait up to 80 seconds for a more delicious treat of a cashew. "The cockatoos held the reward in their beaks directly against their taste organs during the entire delay," says Auersperg. This requires some very impressive self-control. (Imagine a child holding a raisin on her tongue while she waits for a piece of chocolate.) Crows will wait up to several minutes for a better treat. However, if the delay is more than a few seconds long, they'll place the first reward out of sight while they're waiting.

"They do this because they're food cachers, and that's an important part of their ecology," explains Auersperg. Deciding to delay gratification requires not only self-control but also the capacity to assess a respective gain in the quality of a reward relative to the cost of waiting for it—not to mention the reliability of the individual doling out the rewards. These kinds of abilities, thought to be the precursors of economic decision making, are rare in nonhumans.

Ravens have a remarkable ability to remember relationships. Young ravens belong to so-called fission-fusion societies. Before they settle down to a paired, territorial life, they hang out in social groups, where they form valuable alliances with friends and family members. They pick special individuals to share food with, to sit close to (within reach of the other's beak), to preen with and play with. But unlike stable chicken flocks, the social groupings of ravens shift, break apart, and come back together over seasons and years. So the birds face the challenge of keeping track of individuals that are coming and going. Do they recall their affiliations after long periods of separation?

Thomas Bugnyar, a cognitive biologist at the University of Vienna, recently sought to answer this question in a study of a social group of sixteen young ravens in the Austrian Alps. As far as scientists knew, a bird's long-term social memory was limited to remembering neighbors from one breeding season to the next. But Bugnyar found that ravens remember their valued friends even after a separation of as long as three years.

It's worth noting that corvids recognize and recall not only fellow corvids but humans, too. They can pick out familiar human faces from a crowd, particularly those that represent a threat—and remember them for long periods of time. Just ask Bernd Heinrich, who has tried to conceal his identity from the ravens he works with by changing clothing; wearing kimonos, wigs, and sunglasses; and hopping or limping to shift his gait. (The birds weren't fooled.) Or John Marzluff, who describes walking across the campus of the University of Washington and being

singled out from thousands of other people by American crows that recognize him as a dangerous person who has trapped and banded them. The
disgruntled crows still remember him years later and harass and scold him
whenever they spot him. In a brain-imaging study on the crows, Marzluff
recently discovered that the birds recognize human faces using the same
visual and neural pathways that we do.

Pinyon jays use impressive social reasoning to figure out where they
fit in their flock's social order. Avid socialites of the crow family, these
jays live in large permanent flocks with firm social hierarchies, like
chickens. They depend on an understanding of third-party relationships
to work out how to behave toward an unfamiliar jay, whether to be
aggressive or submissive. Think of it this way: A strange jay (we'll call
him Sylvester) enters your flock. It's clear that your flock mate Pete dominates over Sylvester. And you know that Henry dominates over Pete.
Who is more dominant, Henry or Sylvester? Pinyon jays can infer the
social status of a stranger by the way it behaves with other birds, thereby
avoiding unnecessary conflict—and possible injury. This ability to make
judgments about relationships on the basis of indirect evidence is called
transitive inference and is considered an advanced social skill.

I'M FOND OF JAYS, so brash, quarrelsome, jeering. The flocks of blue
jays (*Cyanocitta cristata*) in my region are known for their close family
bonds and complicated social systems, as well as their keen intelligence
and sweet tooth for acorns. They have a way of exploding on the scene,
shrieking at one another, chaffing, scoffing, scolding, barking "like blue
terriers," as Emily Dickinson said. Blue jays can select fertile acorns with
88 percent accuracy. They can also count to at least five. And they can
neatly mimic the piercing cry of a red-shouldered hawk, *kee-ah, kee-ah*—
which they do often, perhaps to fool other birds into believing there's a
raptor in the vicinity, leaving more nuts for the taking. No wonder Bluejay
is the trickster hero of the Chinook and other Northwest Coast tribes.

One species of Old World jay displays especially endearing social acumen. A colorful member of the intelligent crow family, the male Eurasian jay appears to intuit his mate's state of mind—or at least her appetite—and responds by giving her what she most desires.

The jay's Latin name, *Garrulus glandarius*, would seem to explain everything. Eurasian jays are chatty. But they're actually not as gregarious as their more communal cousins, rooks and jackdaws, which nest in crowded rookeries. Pair-bonding is their thing.

Like many other corvids, Eurasian jays share food, but they do so only to win the favor of their mates. A male courts a mate by selecting tasty gifts for her. Ljerka Ostojić and her colleagues at the University of Cambridge recently used this specialized form of gift giving to probe whether these birds might be able to understand that other birds (in this case, their mates) have their own needs and desires, a sophisticated social ability called state attribution.

In an elegant experiment, male jays were allowed to watch through a screen while their mates ate their fill of one of two special treats, wax worms or mealworms. (These goodies might not sound tasty to you, but wax worms are the "dark chocolate" of the jay world.) The males were then given a choice of what to offer as a larval gift—wax worm or mealworm.

Birds, like people, favor variety and can fill up on too much of a good thing. It's called the specific satiety effect. (You know the feeling. You've been gorging on cheese—couldn't eat another piece—so you switch to fruit.) A female jay's penchant shifts with her experience. It behooves a male to track these varying preferences, as giving his mate the food she most desires strengthens his bond with her. Sure enough, when a male jay was able to see his lady's choice in feasting in this trial, he chose to offer her the treat she hadn't been eating.

But maybe he was just considering what might taste good himself. If watching her eat wax worm larvae diminishes his own appetite for that delicacy, this might govern his choice of what to offer her next. However, it turns out that watching her feed on one dish or the other has no effect on what he selects to eat himself. When there is no opportunity to feed

his mate, he chooses between the two foods according to his own preferences. When he can share with her, he disengages from his personal
desires and anticipates hers, as if he's cognizant of her specific satiety. He
offers her the morsel of her choice with the same gentle courtesy a squire
might in serving his lady a slice of her favorite chocolate cake.

This may not be exactly like human state attribution, the ability to
infer that others possess an internal life similar to—but different from—
one's own. Still, it seems pretty close. The Eurasian jay demonstrates
that he can understand his mate's specific desire state. (She wants this,
not that.) He can understand that it differs from his own. (I may have just
eaten a wax worm, but she did not.) And he can (and will!) flexibly adjust
his food-sharing behavior to suit her particular desires.

"These experiments provide exciting data that are in line with the
notion that the male is attributing a desire to his female partner," says
Ostojić. "However, we need to conduct further studies to illuminate
exactly what cues the males are using to respond to the female's specific
satiety. We need to disentangle whether the male is responding purely
to observable features of the female's behavior or whether he can use
those observable features to infer the female's desire."

The possibility that a male jay may be able to intuit his mate's appetites
by watching her hints at the possibility that birds may possess a key component of what's known as theory of mind, the understanding that others
have beliefs, desires, and perspectives that are different from one's own.

"Attributing desires to others is cognitively less demanding than attributing beliefs," says Ostojić. "For humans this is an early step toward
the development of a full theory of mind. If the male jay really understands what the female wants, this would provide evidence that a nonhuman animal is capable of this important aspect of theory of mind."

Ask a dozen experts in the field of animal cognition whether nonhuman animals have theory of mind and you'll get a dozen different
answers. Generally there are two camps: first, the self-described killjoys,
who deny that nonhuman species have anything remotely resembling this
kind of advanced cognition; and second, those who echo Darwin's claim

that humans differ mentally from other species only in degree but not in kind. Two scientists at the University of Pennsylvania, Robert Seyfarth and Dorothy Cheney, lodge in the latter camp. They argue that even the most complex human forms of theory of mind have their roots in what they call a subconscious appreciation of others' intentions and perspectives. At the very least, the scrub jays appear to possess these building blocks of theory of mind.

THERE ARE BIG PAYOFFS to being social: more eyes to spot predators and locate food and plenty of opportunities to learn from others. This means you don't have to waste time figuring out how to open a nut, or risk eating a poisonous berry. You can imitate good ideas and follow flock members to the richest, safest food sources. Rooks and ravens, for example, rely on other flock members to find fruitful foraging patches, massing together around especially rich ones.

According to Lucy Aplin, tits use their social connections to locate food and to copy strategies for getting it, transferring information from flock to flock and even across species. A researcher at the University of Oxford, Aplin studies the social nature of those big tit flocks at Wytham Woods. To investigate the birds' social networks and associations (their version of Facebook), Aplin and her colleagues fitted the tits with tiny electronic tags that tracked their visits to a grid of feeding stations. At the same time, the team evaluated each individual bird's personality with a test that measured their boldness and exploratory behavior.

It should be noted here that birds do have personalities. Some scientists shy away from the term, with its anthropogenic overtones, preferring temperament, coping style, behavioral syndrome. But call it what you will, individual birds behave in ways that are stable and consistent across time and in different circumstances, just as we do. There are the bold and the meek, the curious and the cautious, the calm and the nervous, the fast learners and the slow learners. "The variation in personality is thought to reflect a difference between individuals in their response to risk," explains Aplin.

Scientists recently identified such personality differences in the chickadee, which helps to explain the diverse behaviors around our freshly filled birdfeeder—the one apparently tyrannical little individual that is so adept at hogging all the seed, while another skulks timidly along the sidelines. Some chickadees are bold, "fast" explorers, slapdash and reckless, while others are "slow," cautious and thorough. We take for granted the range of personality differences in our own species. Why shouldn't such variety exist in others?

The study by Aplin's team not only revealed affiliations between birds with similar personalities. It also found that bolder birds flit between groups, expanding the size of their social networks and enhancing their access to information about food sources. "This is especially important in winter, when finding a new good feeding patch might mean the difference between life and death," says Aplin. "However, this behavior may also be a socially 'risky' strategy, increasing the birds' exposure to predation and disease"—which may help to explain why shyness as a trait persists in these birds. The team also found that different species of tits— great, blue, and marsh—share news of food with one another. "The marsh tits are the best information providers," observes Aplin. "They act as a kind of 'keystone' species in the information landscape."

In Sweden and Finland, research revealed that one species may learn from another about not just food but what constitutes good real estate. The experimenters marked nesting boxes with white circles or triangles in an area where tits and migrant flycatchers both nest. Female flycatchers arriving late in the nesting season seemed to cut their risks by choosing to nest only in boxes with the same symbols as those adorning boxes already anointed as choice nesting sites by tits.

In other words, social birds can exploit information offered by other birds. This includes parents and peers and even other bird species. Scientists believe that the pressure to exploit these social sources of information have not only given some birds an advantage in the struggle to survive and reproduce but may also have helped to drive the evolution of their relatively large brains.

BIRDS TURN OUT TO BE very good indeed at learning from their comrades.

Think back to those famous British tits that learned how to open milk bottles in the early twentieth century, a trick one bird learned from the next until, by the 1950s, milk bottles all over England were under siege. To see how this social learning might work, Aplin and her colleagues recently devised an ingenious experiment: They planted new behaviors in the great tit populations at Wytham Woods and watched how they spread.

The team brought a few birds into captivity and trained them to solve a simple foraging puzzle. The birds had to push a sliding door either left or right to gain access to a feeder hidden behind the door. Some birds were trained to push the door to the right; others, to the left. Then all the birds were released back into their woods, which had been seeded with these foraging puzzles. The puzzles were outfitted with specially designed antennae to detect the tiny electronic tags worn by the tits, so that they logged information about visits from every individual bird.

The results were remarkable. The trained birds remained faithful to the side they had been trained on, and within days the researchers saw the same behavior taken up by local birds in each area, with a rapid spread through social network ties to most of the local population. Even if a bird discovered it could push the other side to get the same reward, it stuck with the local tradition. And birds that moved into a new part of the woods from an area with a different bias switched their technique to match the local way of doing things. Birds, like humans, seem to be conformists. A year later, birds remembered their preferred technique, says Aplin, "and the bias still held, even when the behavior spread to a new generation of birds."

This kind of social learning—copying fellow birds in a local environment—say the researchers, might be a quick and cheap way of acquiring successful new behaviors without undertaking potentially risky trial-and-error learning. It is also, says Neeltje Boogert, "the first experimental evidence of persistent cultural variation in new feeding techniques, once thought only to exist among primates."

SOCIAL LEARNING clearly plays a major role in the lives of birds, and not only in the food domain. Female zebra finches learn about mate choice from other females. Say a virgin female sees another female mating with a male wearing a white leg band. Later, when she is presented with two banded but unfamiliar males, one wearing a white band, the other an orange one, she will pick the guy in white.

Then there's the matter of learning to recognize predators or threats. You would think that responding to a predator—such as a raptor or a snake—might be hardwired into a bird. Indeed, some responses are innate. But in spotting novel dangers, copying your confederates comes in handy. One experiment showed that European blackbirds learn to mob a species of bird usually considered innocuous—the Australian honeyeater—after they witness other blackbirds mobbing it.

Birds learn about brood parasites in similar fashion. Young superb fairywrens, for instance, are initially uninterested in the presence of a bronze cuckoo. But after they watch other fairywrens harassing the cuckoos, they change their tune when they see one themselves, emitting whining and alarm calls that instigate mobbing, and setting upon the cuckoos.

A brilliant string of studies over the past five years by John Marzluff and his colleagues at the University of Washington have revealed the extraordinary abilities of American crows not just to recognize individual humans by their faces but to pass along to other crows information about those whom they deem dangerous. In one experiment, teams of people wandered through several Seattle neighborhoods, including the University of Washington campus, wearing different sorts of masks. One type of mask in each group of people represented the "dangerous" mask (on campus, it was a caveman mask). The people wearing the dangerous mask captured several wild crows. The other people, sporting "neutral" masks or no masks at all, just meandered along harmlessly.

Nine years later, the masked scientists returned to the scene of the crime. The crows in these neighborhoods—including those that weren't even hatched at the time of the capture—reacted to the people with the dangerous masks as if they were a threat, dive-bombing, scolding, and mobbing them. Apparently, the birds that witnessed the original capture and those that participated in later mobbings remembered which masks represented danger—and demonstrated this to other crows, including their young. This tendency to mob the dangerous mask spread to crows a half mile or so from the original neighborhood areas, perhaps by way of crow "information networks."

LEARNING BY OBSERVING or imitating is one thing. Learning under the tutelage of a teacher is quite another. More than two hundred years ago, Immanuel Kant argued that "man is the only being who needs education." This view—that teaching is a uniquely human form of social learning—has held on stubbornly. Today skeptics still question whether teaching exists anywhere in the animal kingdom apart from *Homo*. True teaching, the thinking goes, requires all kinds of cognitive abilities that other animals simply don't possess, such as foresight and intentionality, as well as understanding that another being is naïve and other aspects of theory of mind.

But growing evidence suggests that some nonhuman animals do in fact show forms of teaching. Meerkats, for instance, seem to instruct their pups in handling tricky prey, such as snakes or scorpions (which possess neurotoxins potent enough to kill a human). Adult meerkats offer the youngest, most inexperienced pups dead or disabled prey (for instance, a scorpion dispatched with a quick bite to the head or abdomen). But as the pups grow older, their instructors gradually introduce more and more challenging live prey. Giving a naïve pup a wriggling scorpion or a snake that might slither away means that both teacher and pupil may well lose out on a meal. But the effort eventually results in a pup developing skillful hunting and handling of difficult prey. Even ants apparently teach. Scientists have observed experienced tandem-running ants modifying their journeys when trailed by a naïve follower, pausing en route to let a follower-pupil explore landmarks and resuming the journey only when the follower taps them with an antenna.

Still, convincing examples of animal teaching are rare—which is one reason the apparent pedagogy of the pied babbler is so intriguing.

THE SOUTHERN PIED BABBLER (*Turdoides bicolor*) is a striking white bird with dark chocolate flight and tail feathers that thrives in the scrub and savanna of southern Africa. The babblers live in small close-knit family groups of five to fifteen and are highly social and garrulous (not unlike meerkats, those models of mammal sociality). Known in Afrikaans as the "white cat-laughers," they are a noisy lot, named for their constant chattering and group choruses of *chuck-chuck* or *chow-chow-chow*. They never stray far from one another, but forage, preen, play-fight, and huddle together. When one babbler flies, so go the others.

Amanda Ridley, the principal investigator of the Pied Babbler Research Project, studies the birds in the southern Kalahari Desert of South Africa. Babblers are cooperative breeders. Family groups are dominated by a single breeding pair, along with several other adults that don't get to breed, but that nevertheless help to feed and care for the young. The dominant pair is monogamous not only socially but sexually, too—a rare thing in the bird world. In any group, 95 percent of the chicks belong to this pair. Still, all the adults in the group dote on the young, helping to brood, feed, and care for them. If the breeding pair does not produce offspring, babblers have been known to kidnap a young fledgling from another group and raise it as their own.

Babblers spend around 95 percent of their waking time rummaging in the leaf litter for beetles, termites, insect larvae, and burrowing skinks. Foraging with your back to the world is a risky thing for a babbler. Up the food chain and on the prowl for bug-probing birds are the African wild cat and the slender mongoose, the Cape cobra and the puff adder, the spotted eagle owl and the pale chanting goshawk. So dangerous is it for a babbler to have its head down that the birds take turns acting as sentinels, forgoing their own foraging to keep a lookout on behalf of the group for trouble arriving by land or by sky. The sentinel perches in an open spot above the foragers and actively scans for predators, sending up harsh, repetitive peeping alarm calls whenever necessary and offering the group continuous news on its monitoring in the form of a "watchman's song."

Other bird species take clever advantage of the babbler's elaborate sentinel system. Small solitary birds called scimitarbills are known to eavesdrop on the babblers' watchmen. These little "public information parasites" hang around the babblers when they forage, listening in on their alarm calls. This allows the lone scimitarbills to be less vigilant themselves, spending more time foraging in more places, with more success, and even venturing out into the open without worrying about predators. Fork-tailed drongos are more uncouth in their mooching. Highly intelligent, accomplished mimics, they sound false alarm calls of babblers and other species, which make the babblers drop their mealworms and run for cover. Drongos then steal in to seize the dropped food even if it's abandoned only for an instant, right beside the unwitting victim. Ridley and her team recently found that drongos fool the babblers by varying the type of alarm calls they produce, making it harder for the babblers to detect the deception.

Being a sentinel babbler is a dodgy job—they get picked off much more often than foragers, especially by hawks and owls. But life can be dicey for all babblers. Here's where the teaching comes in.

Ridley and her colleague Nichola Raihani have found that a few days before young babblers fledge, adults begin to emit a soft "purr" call when they bring food to the nest, accompanied by a gentle wing flutter. This is the training period: Purr call means food. The adults start to use the call only as the young approach fledgling age. "As the young come to associate the call with food, the adult can then 'bait' them by making a call when holding a food item, but not actually feeding it to them until they have successfully responded to the call," says Ridley. "The young try to reach for it, but the adult moves back out of reach, away from the nest, forcing the nestling to follow. This 'baiting' tactic appears to be a way that parents can 'force' young to fledge"—urgent business, as the chance of nest predation increases as the chicks get older.

After a chick fledges, adults use the special call to move it away from danger and toward good foraging patches. This is more complicated than it sounds. Adults are not teaching their chicks a simple fact, such as the specific location of a foraging site. This would be somewhat useless, as most babbler foraging patches are ephemeral. Rather, they're instructing fledglings in the skill of determining the traits of a fine foraging patch— packed with prey, far from predators. They're also teaching the young how to respond appropriately to a threat by moving them away from unsafe areas when a predator is around, says Ridley. "So the call serves two purposes post-fledging: learning about good foraging patches and learning how to effectively evade predators."

Fledglings, for their part, are not passive pupils. The studies by Ridley and her colleagues suggest that the young birds use at least two clever social strategies to boost the amount of food they get. First, they're picky about whom they follow, choosing to tag along with adults who are especially proficient at capturing prey. Second, when they're hungry, they "blackmail" adults into feeding them at higher rates by venturing into riskier open locations. When they're satiated, they stay in the cover and relative safety of trees.

It remains an open question whether the teaching shown by pied babblers requires sophisticated cognitive abilities. It may be governed by simple processes, such as the more reflexive responses that appear to be part of meerkat teaching. Meerkats may teach their pups by instinctively responding to changes in the pups' begging calls as they grow older: A young pup's call means bring dead prey; an older pup's call, bring live prey. But as Ridley points out, "Teaching pied babblers and teaching in meerkats is different. Meerkats tend to exhibit opportunity teaching (where the teacher puts the pupil in a situation conducive to learning a new skill), while pied babblers tend to display coaching (where the teacher directly alters the behavior of the pupil)," she explains. "We can't rule out entirely that the teaching we see in babblers is the result of reflexive responses—more research is necessary—but it certainly seems like some cognitive abilities are required to carry out this type of coaching behavior."

Ridley suspects that teaching may occur in other bird species with mobile young who accompany foraging adults and use cues from them to find food—such as Arabian babblers, white-winged choughs, Florida scrub jays, and white-browed scrubwrens. "A number of my colleagues have noticed this behavior in the species they study," she says, "so this type of teaching may be more widespread than we currently realize."

SCIENTISTS HAVE FOUND this sort of surprising social ingenuity in the lives of many bird species. What they haven't found is something they expected: a correlation between a bird's social group size and its brain size.

The social intelligence hypothesis predicts that animals living in big social groups will have larger than expected brains because of complicated social pressures. Indeed, when Oxford anthropologist and evolutionary psychologist Robin Dunbar compared brain sizes in different species of primates, he found that those living in larger social groups had bigger brains. In monkeys and apes, the brain size of a species increased in lockstep with the size of the species' group. In primates, group size is considered a proxy for social complexity, which may lead to more advanced cognition.

A clever computer simulation recently offered some virtual evidence for this line of thought. Scientists at Trinity College in Dublin created a computer model with artificial neural networks that served as "minibrains." These minibrains could reproduce. They could also evolve, with random mutations introducing new bits and pieces into their little networks. If these new pieces benefited the network, it would grow in intelligence and could reproduce again, passing along a little goose in brainpower. When the scientists programmed the minibrains to steer through challenging tasks that required cooperation, the minibrains "learned" to work together. As these minibrains got "smarter," cooperation began to accelerate, along with the evolutionary pressure for larger brains. The findings support the idea that complex social interactions like cooperation provided the selection pressures necessary for the evolution of bigger brains and advanced cognitive abilities in our primate ancestors.

However, when Dunbar and his colleagues looked at birds and other animals, the bigger-social-group-equals-bigger-brain pattern didn't hold. The birds with the biggest brains didn't live in large flocks. On the contrary, they favored small cohesive groups and lived mostly in lifelong pairs.

For birds, it seems, the quality of relationships, not the quantity, calls for additional brainpower. The mental challenge is not remembering the individual characteristics of hundreds of individuals in large flocks or roosts or managing a large number of casual relationships. The really demanding task—at least from a psychological and cognitive point of view—is forming close alliances, especially forging bonds with a mate and providing long-term parental care to young.

WE ALL KNOW the challenges: conferring, consulting, coordinating, compromising, factoring in a mate's needs in planning the day.

It's similar for many birds.

About 80 percent of bird species live in socially monogamous pairs, that is, they stay with the same partner for a single breeding season or longer. (That's in stark contrast to the roughly 3 percent of mammal species that exhibit this sort of social monogamy.) This is largely because the business of feeding nestlings is so taxing, requiring biparental care. Birds with altricial young, especially, work their tail feathers off to feed them. Without the contributions of both male and female, few altricial offspring would make it to the fledgling stage. It makes sense to share the burden. But doing so—jointly incubating eggs and feeding and protecting the young—requires careful coordination and synchronization of activities. And that means being tuned in to a mate's little quirks, wants, and needs, and day-to-day shifts in behavior.

According to cognitive biologist Nathan Emery, being bound up with one partner in this way requires a special form of cognition. Called relationship intelligence, it's the ability to read a partner's subtle social signals, respond appropriately, and use this information to predict his or her behavior. And it takes considerable mental acumen.

Some birds reinforce their bonds through fancy acts of coordinated body movements or vocalizations. Rook pairs, for instance, join in a tightly synchronized display of bowing and tail fanning. Plain-tailed wrens, shy, drab little birds living deep in the cloud forests of the Andes, sing rapidly alternating syllables so perfectly coordinated that it sounds like a single bird singing alone. Their duets are a kind of sophisticated auditory tango, demonstrating a truly astonishing level of cooperative behavior. The birds in a pair can sing alone, but when they do, they leave longer gaps between song syllables, in which their partner normally interjects a brief note. This suggests that each member of a pair knows its part in the song, but also relies on auditory cues from its partner to determine when and how to sing. It's a lot like the give-and-take of conversation. Performing duets with such high coordination requires being closely "tuned in" to your mate—and thus may communicate the strength of the pair-bond and level of commitment to each other.

A male budgerigar (*Melopsittacus undulatus*) shows his commitment to his mate by drumming up a perfect imitation of her "contact" call, the special call she uses to keep in touch with her partner as she flies, feeds, and otherwise goes about her day. These small sociable Australian parrots are monogamous but also very gregarious; they like to hang out in large flocks. After only a few days together, pair-bonded budgerigars can converge on the same contact call, with the male managing a bona fide imitation of the female. Her call becomes his. The female uses the accuracy of his imitation to judge his commitment to courting her and his suitability as a mate. Nancy Burley of the University of California, Irvine, and her colleagues who study the budgerigar suspect that this may be the evolutionary reason for the ability of parrots to parrot—to quickly learn and mimic new sounds: "It could also explain why parrot enthusiasts suggest that the 'best talkers' among pet budgerigars are typically males that were obtained when very young and kept in isolation from other budgerigars," write the scientists. "Budgerigars raised under these conditions probably become imprinted on humans and may begin to court them."

WHAT'S ACTUALLY GOING ON in a bird's brain when it's being sociable? Why do some birds form powerful pair-bonds and not others? And why are some kinds of birds loners and others social butterflies?

The late James Goodson looked deep into the brains of birds to try to answer these questions. A biologist at Indiana University until his untimely death from cancer in 2014, Goodson studied the neural circuitry of social grouping in birds. He was interested in understanding the brain mechanisms that determine how birds make social decisions about whom to hang out with and in what size flock.

According to Goodson, the circuits in the brains of birds that control social behavior are much like the circuits in our own brains. The circuits are old—so old they are common to all vertebrates, dating back something like 450 million years, to the common ancestor of birds, mammals, and sharks. The neurons that form them respond to a group of evolutionarily ancient molecules called nonapeptides. The original role of these molecules was to regulate egg laying in our ancient bilaterally symmetrical ancestors (creatures known as bilaterians), but other social functions have evolved from this. In birds, Goodson found that differences in social behavior are rooted in subtle variations in the expression of the genes for these molecules. The same is likely true for humans, too.

In our brains, the nonapeptides are known as oxytocin and vasopressin. Oxytocin, which is made in the hypothalamus of the brain, has been dubbed the love chemical; the cuddle, or trust, hormone; and even the moral molecule. In mammals, it plays a key role in giving birth, lactating, and maternal bonding. In the early 1990s, neuroendocrinologist Sue Carter added pair-bonding to oxytocin's résumé. She and others discovered that prairie voles, which pair for life, have higher levels of the molecule compared with other vole species that are promiscuous.

New research shows that food sharing in chimps raises oxytocin levels more than grooming does. This is evidence, perhaps, for the truth of the maxim "The way to your lover's heart is through her stomach" (and perhaps a window on the Eurasian jay's attention to his mate's appetites).

In humans, oxytocin has been shown to reduce anxiety and promote trust, empathy, and sensitivity. Recent studies have suggested, for instance, that a dose of oxytocin administered through the nose boosts the cooperation of sports team members and makes people more generous and trusting in role-playing games. It also may contribute to the strength of romantic bonds for men by enhancing their brains' reward response to the attractiveness of their partner compared with other women.

Birds have their own versions of these neurohormones, called mesotocin and vasotocin. Over the past several years, Goodson and his colleague Marcy Kingsbury and their team have explored the action of these peptides in various species of birds that differ in their group size.

Consider the zebra finch, a small, gregarious songbird that normally cozies up to its mate and mingles in flocks of hundreds. The biologists discovered that if they blocked the action of mesotocin in their brains, the birds spent less time with their partners and familiar cage mates and avoided big groups. On the other hand, birds that were given mesotocin instead of the blocker became more sociable and sought more close contact with their partners and cage mates and with larger groups.

Goodson decided to map the receptors for these peptides in the brains of bird species with different group-size preferences (big versus small). Maybe the density and distribution of the receptors were the key to why some bird species were more sociable than others. He focused on the estrildids, a large family of 132 finches, waxbills, and munias. All of these birds have similar ecological lifestyles and mating behaviors. They're monogamous, mate for life, and care for their young together. However, they vary widely in the group sizes they prefer. Goodson trekked as far as South Africa to collect three species of estrildids: two that were reclusive, hanging out only in pairs—the melba finch and the violet-eared waxbill; and one that was "moderately" social—the Angolan blue waxbill. To round out the medley, he added two highly gregarious birds that breed in large colonies—the zebra finch and the spice finch, a handsome chestnut bird from tropical Asia that prefers the company of thousands (and known in one lab as the "hippie," or "pacifist," of finches because it has not been observed to show aggression of any kind).

When Goodson mapped the oxytocin-like receptors in the brains of these birds, sure enough, he found startling differences. The highly social, flocking zebra finches and spice finches had far more mesotocin receptors in the dorsal lateral septum—a key part of the brain involved in social behavior—than did their more solitary relatives.

Curious about whether the oxytocin-like peptides also played a major part in the pair-bonding of birds, Goodson and his colleague James Klatt once again peered into the brains of zebra finches.

You can tell that a zebra finch pair has bonded when the two birds "clump," or perch side by side, follow each other, preen each other, and sit together in their nest. When the scientists blocked the action of the peptides in the zebra finch brain, they found that the birds did not display this normal pair-bonding activity. Only with the peptides active in their brains, apparently, will the birds properly partner up.

Some research suggests that oxytocin may play a similar role in humans. In one study, psychologist Ruth Feldman of Bar-Ilan University in Israel found that levels of the hormone in humans are correlated with the longevity of relationships—couples with more oxytocin have longer-lasting relationships.

However, as Marcy Kingsbury points out, the view of oxytocin in humans and its equivalent in birds as a simple "cuddle molecule" is evolving. Recent studies in finches suggest that the so-called love hormones "can actually mediate aggression and even impair pair-bonding," says Kingsbury, depending on the situation. Whether that applies to humans, too, remains to be determined, but in the view of Kingsbury and her colleagues, it seems likely given the similarities in the anatomy and function of these hormones across vertebrate classes. Indeed, some studies of human couples show the opposite of what you might expect: correlations between oxytocin and negative emotions such as anxiety and distrust.

Kingsbury and others argue that there are no neurochemicals that have exclusively "good" or prosocial effects on the brain and body. When it comes to the social effects of these hormones, it seems, context and individual differences matter in both birds and humans.

IN ANY CASE, even paired birds with their cuddle hormones up and running are not paragons of fidelity. According to Rhiannon West, a biologist at the University of New Mexico, this may be another reason why some bird species are smart. West proposes that it's not just the challenges of maintaining pair-bonds in birds that have boosted their brainpower. Rather, she says, it's "the complexity of achieving a successful pair bond *and* extra-pair copulations that is simultaneously driving the increase." It's what she calls an "intersexual arms race."

A few decades ago, science considered birds the very models of sexual monogamy. In the Nora Ephron film *Heartburn*, the female lead bemoans her husband's philandering, and her father responds, "You want monogamy? Marry a swan." But thanks to years of field observations and the advent of molecular "fingerprinting," we now know that swans aren't sexually exclusive and neither are most other birds. DNA analysis has revealed that extra-pair copulations occur in about 90 percent of bird species. In any given nest, up to 70 percent of chicks are not sired by the male caring for them. Pair-bonded birds may be socially monogamous, but they're rarely sexually (or therefore genetically) monogamous. If West is right, this, too, may be a driving force in the evolution of enhanced brainpower.

Take the Eurasian skylark (*Alauda arvensis*), an Old World lark that lives in open grasslands, marshes, and heathlands across Europe and Asia and is noted for singing extraordinarily long and complex songs of up to seven hundred different syllables on the wing. Skylarks are generally socially monogamous. Though the male doesn't help build the nest or incubate the eggs, he does contribute up to half of the food for the nestlings and even more after the chicks have fledged. However, scientists found that 20 percent of the offspring in skylark nests were not genetically related to the male bird attending the nest.

It's easy to see how males would benefit from promiscuity. More liaisons mean more offspring. But what about females? If a male's share of paternity became too low, he might withdraw his parental care. Why would females risk that?

There's no shortage of theories. The going notion is that a female copulates with other males to increase the genetic diversity of the brood (which presumably would boost their chances of survival as long as the male providing parenting services didn't find out), or, perhaps, to garner better genes than those furnished by her partner.

Behavioral ecologist Judy Stamps has offered another hypothesis for why females seek outside liaisons. Her "re-pairing hypothesis," a kind of divorce-remarriage scenario, suggests that trysting females may be checking out the home territory and parenting skills of other males. If a well-heeled male loses or drops his partner and seeks to replace his old mate, he might well turn to this now-familiar up-and-coming female. By trysting with the male, the female not only secures a first-in-line position with him but gleans information about his potential as a superior parent or partner and the quality of his real estate.

A new theory offered by two biologists from the University of Norway suggests that philandering females are encouraging better cooperation in the whole neighborhood. "Females benefit because extra-pair paternity incentivizes males to shift focus from a single brood towards the entire neighbourhood, as they are likely to have offspring there." This might have several positive effects, the researchers suggest, including less territorial aggression and better group protection from predators. (These findings echo earlier studies on western red-winged blackbirds suggesting that females suffered less predation of their nests when they contained extra-pair young, presumably because the genetic sires participated in defending the nests. There was also less starvation among the young birds in those nests.) In essence, by not putting all their eggs in one basket, so to speak, females are pumping up the public good, encouraging safer and more productive neighborhoods. "Where maternity certainty makes females care for offspring at home, paternity uncertainty and a potential for offspring in several broods make males invest in communal benefits and public goods," say the Norwegian scientists. In other words, what's good for the goose is good for all the local geese and ganders.

As evolutionary biologist Nancy Burley points out, it's unlikely that there's any unitary explanation for extra-pair paternity. "The reason that female birds engage in extra-pair copulations probably varies greatly among species," she says. "And within a species, the decision must reflect individual circumstances."

IN ANY CASE, it's clear that both male and female birds stray. But both also work hard to maintain a bond with their social partner to raise their young. In Rhiannon West's view, this dual life may be a key to the large brain size of socially monogamous birds. Regularly procuring extra-pair copulations while maintaining a social partnership makes for a complex social life—and, in West's view, an intersexual cognitive arms race.

Think of it. A male faces the neurological demands of sneaking off to copulate with other females while still actively guarding his mate to minimize cuckoldry. To reduce the chances of an intruder winging in for an extra-pair copulation, for instance, a male skylark must guard the nest closely before his mate lays her eggs. However, he also has another vital job—protecting his territory. So even when he's guarding his mate, he continues to perform the astonishing aerial song flights that serve as a territorial flag signaling "This place is mine." These aerial displays of flapping, gliding, circling, and plunging can last many minutes and typically take place at altitudes of more than six hundred feet. It takes some fancy maneuvering to guard both mate and territory—and still perhaps have time and opportunity for a tryst of his own.

The female, for her part, requires her own suite of cognitive capabilities, not just to sneak off for her own rendezvous but to assess potential partners for genes or real estate—not to mention sustaining the spatial memory to find her way back to her chosen mate. Indeed, in species with more extra-pair paternity, females have relatively larger brains than males; the reverse is true in species with less extra-pair paternity.

The result of all this philandering in birds while at the same time tending to long-term pair-bonds? A boost in brain size for both sexes.

THERE MAY BE ANOTHER social arms race goosing bird intelligence. This one involves the pilfering of food, not sex.

Jays again. This time, the western scrub jay, *Aphelocoma californica*. True to its common name, the saucy scrub jay is a commanding presence in the scrublands of the open West. It scopes out its territory in agile hops and bold lunges, pumping its tail and looking around with quick twists of the head. It misses little. Azure like its blue jay cousins (though without the jaunty crest), it is equally impudent, known as a thief, scoundrel, and jackal of the bush. According to one ornithologist, a favorite trick of the jay is to rob a cat of its food by giving its tail a vigorous peck and, "when the cat turns to retaliate, to jump for the prize and make off with shrieks of exultation."

The scrub jay lives in monogamous pairs throughout the year, often within flocks. But during the breeding season, each male bird acts as if he owns the place, staunchly defending his territory from rival scrub jays with darting flights and shrieking calls. "The jay's ordinary alarm note is an astonishing vocal outbreak, *dʒweep, dʒweep*, with which the groves are brought up standing," writes one naturalist. "It curdles the blood, as it is meant to do."

Scrub jays are a caching species. All autumn long, they dart about the underbrush gathering acorns and other nuts by the thousands, as well as insects and worms. These they hoard for future use in thousands of caches throughout their territory.

It all sounds very honorable and industrious, except for one thing. The birds live a kind of double life, storing their own food for future use while raiding the stores of other birds. They are cachers, all right, but they are also thieves who mine the hard-won booty of their neighbors.

A scrub jay may lose up to 30 percent of his stash of stored food per day—no small thing for a bird that must squirrel away enough food to weather long, harsh winters. Pilferage, the loss of cached food items, is a big problem, plainly one of the drawbacks of social living.

But there's an interesting twist. The interaction between food cachers and food thieves in the scrub jay community appears to have led to the evolution of some astonishingly smart behavior—tactical deception, on the part of both the food storer (to protect its caches) and the potential pilferer (to outwit the cacher and competitive pilferers to win the caches).

In a series of inspired studies, Nicola Clayton and her colleagues have found that scrub jays will go to great lengths to protect information about the location of their caches from pilferers. A caching scrub jay will opt to hide his food behind a barrier or in the shade over a more obvious, well-lit place in the open—if and only if another bird is watching him. (When the observing bird's view is blocked, the jay won't bother trying to cache in a more private place.) If the observer can hear him, but not see him, he'll cache his food in a less noisy substrate—in quiet soil instead of pebbles. Moreover, if another bird has seen him hide his food in a particular cache, he may return to that spot and move—or pretend to move—the contents of that cache to another spot, in a kind of shell game that confuses the potential thief. He'll even pretend to cache by probing in a new place after the food has already been hidden elsewhere, confusing the pilferer so he can't keep track of where the food ends up. Is there a clearer example of sly trickery?

Not just any observer will spur him to these elaborate tactical strategies. If his mate is watching, he's likely to be perfectly open in his actions. And only if a rival bird has watched him cache in a particular place is he seen as a threat. Somehow, scrub jays keep track of who has been watching, both where and when. They recall whether or not they were observed during a specific caching event—and by whom—and will recache later only if absolutely necessary.

But here's the really amazing thing. A scrub jay will think to do this—to resort to these clever cache-protection tactics—only if he's had his own piratical experience. Birds that have never pilfered themselves hardly ever recache. In other words, say the researchers, "it takes a thief to know a thief."

Pilferers, for their part, try to keep a low profile, hiding while they watch a caching bird and keeping quiet, reducing the chances that the caching bird will think to use one of its cache-protection schemes.

The upshot is a kind of "war of information," with pilfering jays developing strategies for actively seeking information while remaining unobserved, and cachers becoming more and more skilled at developing Machiavellian tactics for fending them off, concealing information, or providing false information.

For Clayton and many others who study the scrub jay, the deceptive and manipulative behaviors of the birds suggest some highly sophisticated thought processes: memory for who was around when and where (known as episodic-like memory), as well as the ability to use the personal experience of being a thief to predict a thief's expected actions, and perhaps even perspective taking—imagining another bird's point of view (what it knows and doesn't know) and tailoring reactions accordingly. The capacity to take the perspective of others—to grasp what might be going on in another creature's head—is one of the hallmarks of theory of mind.

It's not clear whether caching and pilfering have driven the evolution of these cerebral skills. It could also be that the skills already existed in these jays (perhaps as a result of attending to a mate?), and they just applied them to caching. It's the proverbial chicken-and-egg situation, like the crow and its tools.

DO BIRDS EXPERIENCE SUCH prized human social or emotional capacities as empathy or grief? It's a lingering question. As Clayton and her colleague Nathan Emery warn, "For birds, especially those known to be smart such as crows and parrots, it is very easy to slip into the anthropomorphic trap and attribute them with human emotions without good evidence."

But consider the case of the greylag goose (*Anser anser*). A European bird of modest wit, the greylag was made famous by Nobel laureate Konrad Lorenz, who demonstrated that the young birds imprint on anything that moves. Case in point: The goslings he raised by hand followed—and then later tried to mate with—his Wellington boots. Greylag geese live in groups from small families to flocks of thousands and have a social life comparable to brainier birds such as crows and parrots. They show off their social bonds with partners and family members by sticking close and by performing together in a "triumph ceremony," a series of ritualized movements and vocal display. A recent study at the Konrad Lorenz Research Station in Austria measured the heart rate of these geese—a concrete measure of distress—in response to various events: thunder, passing vehicles, the departing or landing of flocks, and, finally, social conflicts. The biggest boost in heart rate, it turned out, occurred not in response to something surprising or frightening, such as a clap of thunder or the roar of traffic, but in reaction to a social conflict involving either a partner or a family member. For the scientists, this points to emotional involvement, possibly even empathy.

Then there is the kissing of rooks. These supremely social members of the crow family nest in crowded rookeries where there are lots of opportunities for tiffs. One study revealed that after watching a partner in a conflict, rooks often comfort the distressed bird within a minute or two by twining bills with it. This was heralded by researchers—albeit in somewhat cold-potato terms—as a triumph of "postconflict third-party affiliation," meaning that after a fight, an uninvolved bystander (third party) offered this tender reassurance to the victim of aggression in the conflict, usually a mate.

Only a few animal species have been known to reassure others in distress, among them great apes and dogs. Asian elephants were lately added to the list with a study showing that they may console a distraught individual with their trunks, gently touching its face or putting their trunk in its mouth—akin to an elephant hug.

Not long ago, Thomas Bugnyar and his colleague Orlaith Fraser set out to discover whether ravens provide this kind of comfort to distressed mates or friends who are victims of a conflict. Do ravens feel sympathy for victims after an aggressive conflict? Do they console them?

Consolation is of special interest, say the researchers, "because it implies a cognitively demanding degree of empathy, known in humans as 'sympathetic concern.'" Consoling a victim means first recognizing suffering and then responding in a way that alleviates it for the sufferer. This requires sensitivity to the emotional needs of others—a trait once thought unique to humans and their closest relatives, chimps and bonobos.

The scientists studied a group of thirteen young ravens. Before young ravens pair up and get territorial, they hang out in large flocks and cultivate valuable allies and partnerships. In any social group, conflicts may arise, and an "unkindness" of young ravens is no exception. Raven fights, especially those within a family, are usually minor squabbles involving a few pecks here and there. But fights between strangers or members of other families over nests, mates, food, or territory can be prolonged and deadly.

Over a two-year period, the researchers carefully observed 152 fights among the young ravens, recording the identities of the aggressor, the victim, and the bystanders—flock members standing near enough to witness the conflict. They rated the fights as mild (mostly noisy threats) or intense (chasing or jumping at another bird or hitting it hard with the bill).

Then, for ten minutes after each fight, they noted any acts of aggression or its opposite, affiliation, with the victims. To their surprise, the researchers found that within two minutes following an intense fight, bystanding flock members offered consoling gestures to the victim of a conflict. The gestures, offered most often by a partner or an ally, included sitting side by side with the victim, preening it, bill twining, or touching its body gently with the bill while uttering soft, low "comfort" sounds. The killjoy explanation: The birds may simply be trying to reduce the external signs of stress in their partner or ally. But to the study's authors, the ravens' comforting behavior appears to arise from knowledge of the feelings of others. These findings, they write, are "an important step towards understanding how ravens manage their social relationships and balance the costs of group-living. Furthermore, they suggest that ravens may be responsive to the emotional needs of others."

AS FOR GRIEF: When the news broke recently that scientists had witnessed a western scrub jay "funeral," my mind went directly to an incident I witnessed years ago in a meadow not far from my home: a flock of blue jays gathering around a red-tailed hawk that had just picked off one of their own. The victim was flailing in the hawk's talons. The surrounding jays were squalling and mobbing the murderer, who seemed unperturbed by their commotion. I hung around for a while until the hawk finally flew off with its plunder, now limp.

But this "funeral" was different. It was created by Teresa Iglesias and her colleagues at the University of California, Davis, who were interested in how scrub jays might respond to the presence of a fellow jay already dead. The team set out a dead jay in a spot in a residential neighborhood where the jays normally forage and recorded what happened next. The first jay to encounter the dead bird responded by calling in other jays with a bloodcurdling alarm call. The jays nearby stopped foraging and flew to the site, joining in a loud, cacophonous gathering, which got bigger and noisier over time.

Were they mourning a fallen member of the jay tribe? Jeering in outrage? Trading ideas about what killed him or how to get him out of this place? The birds congregated around the body for half an hour before they flew away; for a day or two thereafter, they avoided feeding in the area.

Reactions to the study quickly cycled from wonder (birds mourning!) to a heated debate and criticism of the researchers' inappropriate use of the "f-word," as one commentator put it. Some critics smelled straight-out anthropomorphism. This was hardly a funeral in the human sense.

No, but the researchers weren't suggesting that. They were just demonstrating how birds respond to a dead member of their own species: apparently, by noisily telling other birds about the death and perhaps alerting the group to danger, a behavior the scientists called "cacophonous aggregation."

In this sense, perhaps the scrub jay gathering was more like the Irish wake that naturalist Laura Erickson recalled when she heard about this research. The wake was in honor of Erickson's father, a Chicago firefighter who died of a sudden heart attack immediately following a fire. Erickson describes how her father's fellow firefighters streamed in to see him one last time and "talked about how good he looked except for being dead" or "how they should spend more time in the gym or going on a diet of something—the subtext being that they wanted to avoid the same fate."

In a follow-up study, Iglesias and her colleagues found that scrub jays respond with group cacophony when they see dead birds of different species that are about their own size—pigeons, for instance, or American robins or mockingbirds. (In the study, the team used pigeons and two species unfamiliar to the scrub jays, the blue-tailed bee-eater and the black-naped fruit dove.) The jays respond only weakly or not at all to the deaths of smaller species, such as finches. This suggests that these gatherings are used in assessing risk rather than for mourning, says Iglesias. Similarly sized birds tend to share predators. "However," she adds, "this does not preclude the possibility that western scrub jays experience emotional pain during some if not all cacophonous gatherings."

I'M NOT SURE WHAT to make of the scrub jay case. One definition of empathy is "transforming another person's misfortune into one's own feeling of distress." Were those birds in the California experiment simply issuing a warning? Or were they feeling something on behalf of their comrade? Indignation? Fear? Sorrow? Birds may not express emotions through the facial musculature, as primates do, but they can do so using their heads and bodies or through vocalizations, gestures, and displays. Konrad Lorenz once noted that a greylag goose that had lost its partner showed symptoms of grief similar to that of young children who have suffered loss, "the eyes sink deep into their sockets . . . the individual has an overall drooping experience, literally letting the head hang."

The jury is still out on whether birds grieve their own. But more and more scientists seem willing to admit the possibility.

Marc Bekoff, professor emeritus at the University of Colorado, relates a story told by Vincent Hagel, former president of the Whidbey Audubon Society. While visiting at a friend's house, Hagel looked out the kitchen window and saw a dead crow just a few feet away. "Twelve other crows were hopping in a circle around the body," said Hagel. "After a minute or two, one crow flew off for a few seconds, then returned with a small twig or piece of dried grass. It dropped the twig on the body, then flew away. Then, one by one, the other crows each left briefly, one at a time, and returned to drop grass or a twig on the body, then fly off until all were gone, and the body lay alone with twigs lain across it. The entire incident probably lasted four or five minutes."

I've heard other stories like this, of hundreds of crows filling the trees around a golf course after a crow was killed by a golf ball; of a vortex of ravens assembling within minutes at the spot where two ravens roosting on a power transformer were electrocuted. In *Gifts of the Crow*, John Marzluff and Tony Angell suggest that crows and ravens "routinely" gather around their own dead. This response may be more social than emotional, they suggest, as the birds work out what the void means to their group hierarchy, matters of mates and territory, and also, as Iglesias suggests, how they might avoid ending up like their comrade. Marzluff has shown that when crows see a person holding a dead crow, the hippocampus in their brains is activated, indicating that they are learning about the danger. "We are convinced that crows and ravens gather around their dead because it is important to their own survival that they learn the causes and consequences of another crow's death," write Marzluff and Angell. "We also suspect that mates and relatives mourn their loss."

I suspect so, too. Surely grief isn't a human invention any more than love, or deceit, or imagining what your mate might crave for dinner.

five

FOUR HUNDRED TONGUES

Vocal Virtuosity

If you happened to find yourself at the foot of the stairs in the White House on a typical afternoon sometime around 1804 or 1805, you might have noticed a perky bird in a pearl-gray coat ascending the steps behind Thomas Jefferson, hop by hop, as the president retired to his chambers for a siesta.

This was Dick.

Although the president didn't dignify his pet mockingbird with one of the fancy Celtic or Gallic names he gave his horses and sheepdogs—Cucullin, Fingal, Bergère—still it was a favorite pet. "I sincerely congratulate you on the arrival of the Mocking bird," Jefferson wrote to his son-in-law, who had informed him of the advent of the first resident mockingbird. "Learn all the children to venerate it as a superior being in the form of a bird."

Dick may well have been one of the two mockingbirds Jefferson bought in 1803. These were pricier than most pet birds ($10 or $15 then—around $125 now) because their serenades included not only renditions of all the birds of the local woods, but also popular American, Scottish, and French songs.

Not everyone would pick this bird for a friend. Wordsworth called him the "merry mockingbird." Brash, yes. Saucy and animated. But merry? His most common call is a bruising *tschak!*—a kind of unlovely avian expletive that one naturalist described as a cross between a snort of disgust and a hawking of phlegm. But Jefferson adored Dick for his uncommon intelligence, his musicality, and his remarkable ability to mimic. As the president's friend Margaret Bayard Smith wrote, "Whenever he was alone he opened the cage and let the bird fly about the room. After flitting for a while from one object to another, it would alight on his table and regale him with its sweetest notes, or perch on his shoulder and take its food from his lips." When the president napped, Dick would sit on his couch and serenade him with both bird and human tunes.

Jefferson knew Dick was smart. He knew he could mimic other birds in his neighborhood, popular songs of the day, even the creak of the ship's timbers on a crossing to Paris. But what Jefferson could never imagine was how science would come to view the nature of Dick's ability. How rare and risky it is, the brainpower it requires, and how it offers a window into a most mysterious and complex form of learning: imitation, the wellspring for so much of human language and culture.

INSIDE THE LOHRFINK AUDITORIUM at Georgetown University one recent fall day, a flock of 180 specialists have gathered to bandy about new research and ideas about Dick's skill and its parallels with human language learning. The skill is the ability to imitate sounds, to glean acoustic information and use it for one's own vocal production—a vital prerequisite for language. It's called vocal learning, and it's rare in the animal world, thus far found only in parrots, hummingbirds, songbirds, the bellbirds, a few marine mammals (such as dolphins and whales), bats, and one primate—humans.

The specialists are discussing the complex cognition involved in song learning in birds. If cognition is defined as the mechanisms by which a bird acquires, processes, stores, and uses information, then song learning is clearly a cognitive task: A young bird picks up information about how a song should sound by listening to tutors of its own species. It stores this information in its memory and then uses it to shape its own song. The scientists are noting the remarkable similarities of song learning in birds with human speech learning, from the process of imitating and practicing right down to the brain structures involved and the actions of specific genes—how songbirds have "speech defects" just as we do (they stutter, for instance) and the way song learning in a bird literally crystallizes brain structure, teaching us about the neurological nature of our own learning.

Johan Bolhuis, a neurobiologist at Utrecht University, remarks on how strange it must seem to an outsider for scientists to be comparing birdsong with human speech and language. "If we were looking for some kind of animal equivalent, wouldn't we look to our closest relatives, the great apes?" he asks. "But the odd thing is, so many aspects of human speech acquisition are similar to the way that songbirds acquire their songs. In the great apes, there's no equivalent at all."

WHEN I STEP OUTSIDE the auditorium during a break, I notice a small cedar tree—more like a shrub—emanating a bouquet of birdsong. All around campus a cold northwest wind is shaking loose the leaves of oaks and maples and buffeting the occasional swoop of sparrows. Otherwise, there are few birds to be seen. But from the heart of this shrub I catch the *tea-kettle, tea-kettle, tea-kettle, tea-kettle* of a Carolina wren and the gurble of a white-breasted nuthatch. Then the brazen bulletlike *pew pew pew tweeee* of a cardinal and what sounds like the scold of a robin. When I peer between the boughs, I see a single gray bird, feathers puffed against the chill. It's a northern mockingbird (*Mimus polyglottos*, or "many-tongued mimic"), one of Dick's tribe, spilling his soul in song. He pauses for a second or two between each set of phrases, as if mulling over his next selection.

I've seen mockingbirds do this in the peak of spring to stake out their territory and call for mates, mounting to the highest branch to belt out their tunes. One April afternoon, I was at the foot of a solitary pine tree on the flat sandy landscape around the Delaware shore. Unlike the bird in the bush, this one cut a conspicuous figure. Perched erect on the topmost spray of pine, long tail briskly twerking, bill arrowed skyward, he poured out his music with fervor, throwing his whole body into the effort, song after song after song.

The mockingbird is a member of the *Mimidae* family of thrushes found only in the Americas. On the voyage of the *Beagle*, Darwin met with mockingbirds everywhere in South America, where he noted, "They are lively, inquisitive, active birds . . . possessing a song far superior to that of any other bird in the country."

Mockingbirds have been maligned as mere thieves who pilfer tunes and miss the main musical point of their stolen songs. But to my ear, this Delaware bird sang Carolina wren the way Bette Midler does the Andrews Sisters. It may be true that he was a shameless sampler, strewing about phrases from titmouse, chickadee, the sweet liquid song of a wood thrush, but he tucked them into his song the way Shostakovich weaves his symphony around a simple folk melody. After a while, I was lost, so captivated by his choral improvisation that I forgot to listen for familiar songs and calls. His melody filled the warm spring air with countless mimicked swells and trills, joyous and exuberant.

Then, as abruptly as it started, his impassioned outburst ended. He fluttered down from the tree and settled quietly in the leaf litter, as if it just felt good to get the whole thing off his breast.

THAT WAS IN SPRING, when birds sing their hearts out to establish territory or secure a mate. But now it's mid-November, with a chilling wind. This bird is hiding in a cedar tree like a fugitive from justice—singing, it would seem, only to himself. His notes fall into little refrains repeated four or five times, and his songs seem limitless.

How can a bird store so many tunes in a brain a thousand times smaller than mine? And how did those tunes get there in the first place? Why is this bird apparently serenading himself deep inside a bush?

"It's not unlike our singing in the shower," hints Lauren Riters of the University of Wisconsin, one of the birdsong specialists probing such questions in the warmth of the Lohrfink Auditorium.

This bird has spent enormous time and resources learning his oeuvre. Many people assume that birdsongs are genetically encoded. But songbirds go through the same process of vocal learning that people do—they listen to adult exemplars, they experiment, and they practice, honing their skills like children learning a musical instrument.

This is one reason those 180 songbird specialists have grown so deeply interested in their topic. Some of our most complicated skills—language, speech, music—we learn the way birds do, through a similar process of imitation.

"By studying vocal learning in birds, including those that can imitate human speech, such as parrots," says neurobiologist Erich Jarvis, "we can find the essential brain pathways, genes, and behaviors that are necessary for this ability."

ALL BIRDS VOCALIZE. They hoot, yodel, caw, wail, rattle, *chit*, *seet*, and sing like angels. They call to warn of predators and to identify family, friends, and foes. They sing to defend their territory—to stake it out or fence it in—and to woo a mate.

Calls are typically short, simple, succinct, and innate (like a human scream or laughter), uttered by both sexes to make their point. Songs are generally longer, more complex, and learned, sung normally in tropical regions by both males and females, and in temperate climates more commonly by males only during the breeding season. But there is no neat division between call and song, and there are plenty of exceptions. The calls of crows fall into a dozen different categories—rallying, scolding, assembling, begging, announcing, dueting, among others—and some are learned. For sheer complexity, the calls of the black-capped chickadee far and away beat out a great tit's two-tone song.

But singing is something special. "Nearly all animals that communicate vocally do it by instinct," says Jarvis, who studies vocal learning at Duke University. "They are born knowing how to scream or cry or hoot." These utterances are innate or imprinted, like a sheep's baa. "Vocal learning, on the other hand, involves the ability to hear a sound and then, by using muscles of your larynx or syrinx, to actually repeat that sound yourself," explains Jarvis, "whether it be a sound learned in speech or the note of a birdsong."

Close to half the birds on the planet are songbirds, some four thousand species, with songs ranging from the mumbled melancholy chortle of the bluebird to the forty-note aria of the cowbird, the long, byzantine song of the sedge warbler, the flutelike tune of the hermit thrush, and the amazing seamless duets of the male and female plain-tailed wren.

Birds know where to sing and when. In the open, sound travels best a few feet or so above the vegetation, so birds sing from perches to reduce interference. Those singing on the forest floor use tonal sounds and lower frequencies than those singing in the canopy. Some use frequencies that avoid the noise from insects and traffic. Birds living near airports sing their dawn chorus earlier than normal to reduce overlap with the roar of airplanes.

IN HIS POEM "Ode to Bird Watching," Pablo Neruda asks, "How / out of its throat / smaller than a finger / can there fall the waters / of its song?"

Because of a single invention.

It's a unique instrument called a syrinx, after the nymph transformed into a reed by Pan, god of fields, flocks, and fertility. Scientists took a long time figuring out its details because the syrinx is buried deep in a bird's chest, where the trachea splits in two to send air to the bronchi. Only in the past few years did researchers finally produce a stunning high-resolution three-dimensional image of the organ in action, using magnetic resonance imaging and microcomputed tomography.

The high-tech image revealed a remarkable structure. It's made of delicate cartilage and two membranes that vibrate with airflow at super-fast speeds—one on each side of the syrinx—to create two independent sources of sound. Gifted songbirds such as the mockingbird and canary can vibrate each of their two membranes independently, producing two different, harmonically unrelated notes at the same time—a low-frequency sound on the left, a high-frequency sound on the right—and shifting the volume and frequency of each with such breathtaking speed as to produce some of the most acoustically complex and varied vocal sounds in nature. (This is quite extraordinary. When we talk, all of our pitch, all the harmonics of our vocalizations, move in the same direction.)

All of this is managed by minute but powerful muscles. Certain songbirds, such as European starlings and zebra finches, can contract and relax these tiny vocal muscles with submillisecond precision—more than a hundred times faster than the blink of a human eye. This feat of fast-muscle contraction has shown up in only a handful of animals, including the sound-producing organ of rattlesnakes. The winter wren, a little brown bloom of a bird known for its swift song delivery, sings as many as thirty-six notes per second—far too quick for our ears or brain to perceive or absorb. Some birds can even manipulate the syrinx to mimic human speech.

Birds with a more elaborate set of syringeal muscles tend to produce more elaborate songs. The mockingbird in that cedar tree has seven pairs that allow him to perform his vocal gymnastics over and over with seemingly little effort—seventeen, eighteen, nineteen songs per minute when he really gets going. Between the notes, he takes tiny breaths to replenish his air supply.

His phantasmagorical song may be executed by his syrinx, but it's initiated and coordinated by his brain. Nerve signals from an elaborate network of brain areas control each of the muscles, coordinating nerve impulses from his left and right brain hemispheres to the muscles in the two halves of his syrinx, creating just the right airflow in each necessary to produce the hundreds of different imitated phrases he sings.

He makes it all look so easy.

But think about it. To imitate a phrase, say, in German or Portuguese, you have to listen carefully to the person uttering it. You have to hear it accurately. Not such an easy task, Tim Gentner tells the roosting bird specialists at Georgetown, especially if you're at a cocktail party or on a noisy street, where you have to pick it out from a cacophony of sound, a phenomenon called "stream segregation." Birds put up with a lot of this partylike pandemonium, especially at peak times of song, like the dawn chorus. "Many birds are social creatures; they're communicating with one another in relatively large groups," says Gentner, a psychologist at the University of California, San Diego. "There's a lot of signal, and not all of it is useful to every individual at every point in time, so one important task is to figure out which acoustic streams are carrying information."

Once you've segregated a target phrase from all that din, you have to hold it in your mind while your brain translates the sound into a set of motor commands. It sends these to your larynx in the hopes that it will produce a similar sound. Rarely do you get the phrase right the first time. It takes practice, trial and error, hearing your own mistakes and correcting them. If you want to retain the phrase, it means repeating it often enough to reinforce the brain pathways that created the memory in the first place. And if you want to remember it for life, you must file it away in a safe long-term-storage place.

Mockingbirds are really, really good at all this. The proof is in the sonograms or spectrograms. These are visual printouts of sound (with frequency or pitch on the vertical axis and time on the horizontal axis) that scientists use to detect subtle differences in birdsong. Sonograms comparing a prototype song and the mockingbird's copy show that the imitator sings nuthatch and thrush and whip-poor-will with almost perfect fidelity to the original script. Scientists found that when a mockingbird sings a cardinal's song, it actually mimics the muscular patterns of the cardinal. If the notes of his model fall outside his normal frequency range, he substitutes a note or omits it, lengthening other notes to match the song in duration. And if he's facing a too-rapid-fire delivery of notes such as a canary's, he clusters the notes and pauses to breathe while still maintaining identical song length. That may not fool the whip-poor-will or thrush, but it fools me.

Of course, the mockingbird is not the only mimic in birddom. A cousin *Mimidae*, the brown thrasher, by some accounts can mimic ten times the number of songs a mockingbird sings, though not with such accuracy. Common European starlings are also accomplished mimics, as are nightingales, which can imitate some sixty different songs after hearing each only a few times. Marsh warblers are known to sing a wild, urgent, international pastiche of a song peppered with the tunes of more than one hundred other species. Some of the songs are European, picked up at its nesting grounds, but most are African, plucked from the neighborhoods of Uganda, where it spends its winters. Its imitated songs of the Boran cisticola, vinaceous dove, and brubru shrike create a kind of acoustic map of its African travels.

The lyrebird is renowned as a champion sound thief. As one naturalist noted, it's a startling experience to be walking in the Australian forest when suddenly you're confronted "by a fowl-like, brown bird which may bark at you like a dog." The fork-tailed drongo, that brainy African bird that dupes the pied babbler, mimics the alarm calls of not just the babblers but a startling number of other species in a similar ploy—to scare honest birds or mammals off their hard-won morsels, which the drongo then steals.

There are reports of a bullfinch trained to sing "God Save the King," a gray catbird sounding "Taps" (which it may have picked up from burial services at a nearby cemetery), and a crested lark in southern Germany that learned to imitate the four whistling notes a shepherd used to work his herding dogs. So faithful were the imitations that the dogs instantly obeyed the bird's whistled commands, which included "Run ahead!" "Fast!" "Halt!" and "Come here!" These whistled calls subsequently spread to other larks, creating a little pocket of local "catchphrases" (and, quite possibly, some very winded sheepdogs).

Some birds have an exceptional gift for imitating human speech. The African grey parrot is one such species. The mynah certainly qualifies, as does the cockatoo. These few are generally considered the Ciceros and Churchills of birds. Arguably, there are a few others in the corvid and the parrot families: parakeets, for instance. The *New Yorker* once reported that "after weeks of silence, the first words uttered by a Westchester parakeet were, 'Talk, damn you, talk!'"

Imitating human sounds is a lot to ask of a bird. We form vowels and consonants with our lips and our tongue, among the strongest muscles per inch in the human body. For birds, with no lips and with tongues that generally aren't used for making sounds, it's a tall order to take on the nuances of human speech. This may explain why only a handful of species have accomplished the skill. Parrots are unusual in that they use their tongues while calling and can manipulate them to articulate vowel sounds, talents that probably underlie their ability to mimic speech.

The African grey parrot is the parliamentarian of the bird world. Irene Pepperberg made African greys and their speech abilities famous through her work with Alex, perhaps the world's most renowned talking bird. Pepperberg would intermingle different sorts of questions about objects and Alex could answer with near perfect specificity. For example, if she showed him a green wooden square, he could say what color it was, what shape, and, after touching it, what it was made of. He also cottoned to phrases he heard around the lab, like "Pay attention"; "Calm down"; and "'Bye, I'm gonna go eat dinner, I'll see you tomorrow."

Alex was not alone in his badinage. One African grey I know, Throckmorton, pronounces his name with Shakespearean precision. Named for the man who served as an intermediary for Mary, Queen of Scots (and was hanged in 1584 for conspiring against Queen Elizabeth I), Throckmorton has a wide repertoire of household sounds, including the voices of his family members, Karin and Bob, which he uses to his advantage. He calls out Karin's name in a "Bob voice" that Karin describes as spot-on; she can't tell the difference. He also mimics the different rings of Karin's and Bob's cell phones. One of his favorite ploys is to summon Bob from the garage by imitating his cell phone ring. When Bob comes running, Throck-morton "answers" the call in Bob's voice:

"Hello! Uh-huh, uh-huh, uh-huh."

Then he finishes with the flat ring tone of hanging up.

Throckmorton imitates the *glug, glug* sound of Karin drinking water and the slurping sound of Bob trying to cool his hot coffee while he sips it, as well as the bark of the family's former dog, a Jack Russell terrier dead nine years. He has also nailed the bark of the current family pet, a miniature schnauzer, and will join him in a chorus of barking, "making my house sound like a kennel," says Karin. "Again, he's pitch perfect; no one can tell it's a parrot barking and not a dog." Once, when Bob had a cold, Throckmorton added to his corpus the sounds of nose blowing, coughing, and sneezing. And another time, when Bob came home from a business trip with a terrible stomach bug, Throckmorton made sick-to-my-stomach sounds for the next six months.

For one long stretch, his preferred "Bob" word was "Shhhhhhhhiiiit."

Parrots have been known to teach other parrots to talk smack. Not long ago, a naturalist working at the Australian Museum's Search and Discover desk reportedly took a number of calls from people who had heard wild cockatoos swearing in the outback. The ornithologist at the museum speculated that the wild birds had learned from once-domesticated cockatoos and other parrots that had escaped and survived long enough to join a flock and share words they had picked up in captivity—if true, a fine example of cultural transmission.

STILL, THE SHEER PROFUSION and precision of a mockingbird's imitated songs is a marvel. A tally of one mockingbird's tunes captured twenty imitations of calls and songs per minute: nuthatches, kingfisher, northern cardinal, kestrel, even the high-pitched *seep seep seep* begging of a mockingbird chick. The Arnold Arboretum mocker of Boston was said to mimic thirty-nine birdsongs, fifty birdcalls, and the notes of a frog and a cricket. You can tell where a mockingbird lives by the songs he sings.

So particular is a song to its bird that individual birds within a population may share only 10 percent of their song patterns. When it came to describing the mockingbird's imitative skills, the ornithologist Edward Howe Forbush dropped all pretense of scientific detachment, trumpeting the mocker as "the king of song" surpassing "the whole feathered choir." No wonder the Native Americans of South Carolina called the bird Cencontlatolly, or "Four Hundred Tongues." It's only a small exaggeration. Mockingbirds regularly imitate as many as two hundred different songs. Dan Bieker, an ornithologist friend of mine, notes that picking out the imitative songs a mockingbird sings becomes easier over the spring season. "Early in the season their renditions are pathetic, muddled and difficult to identify," he says, "but they get better as they go, as they hear and practice the songs around them—towhee, titmouse, truck backing up, or telephone."

WHY ANY CREATURE WOULD devote so much time and mental energy to imitating other species and random sounds remains a conundrum. Clearly the drongo's mimicry has a very specific purpose. But what about the mockingbird? The fancifully named "Beau Geste" hypothesis suggests that male songbirds flit from perch to perch and sing imitated songs after each move in an effort to make potential rivals think the region is packed with territorial males. The hypothesis is named for the Hollywood film starring Gary Cooper. In that film, Beau Geste bluffs attacking Arabs and protects his fort by propping up his wounded and dead comrades against the fort's parapets and firing his rifle to suggest that every wall is manned with defenders.

Some say vocal mimicry in birds is more like the Batesian version of mimicry: The monarch butterfly mimics the wing design of the toxic viceroy butterfly to warn would-be predators, "Eat me and you'll die." Australian magpies, for instance, mimic nest marauders like the barking owl and the booboo owl—perhaps confusing the owls about the identity of their prey. But this doesn't explain the magpie's mimicry of other sounds.

Or the mockingbird's—which may have more to do with expanding his repertoire to please the ladies. Whatever the motivation, it's an astonishing feat.

AS EARLY AS 350 B.C., Aristotle noted that songbirds learn to sing. "Some small birds do not utter the same voice as their parents when they sing, if they are reared away from home and hear other birds singing." Darwin remarked on it, too. He knew that birds had an instinct to sing just as we have an instinct to speak, but they learn the songs themselves, just as we learn languages. He suspected that birds, like people, pass along their songs from generation to generation, forming regional dialects. But scientists of the 1920s—perhaps under the spell of B. F. Skinner, who thought that many behaviors, even learned ones, were innately determined— decreed that the mockingbird was born with all his song. As ornithologist J. Paul Visscher wrote in the *Wilson Bulletin*, "A Mockingbird does not as a rule consciously mimic songs but only possesses an unusually large series of melodies which it calls forth in wonderful perfection."

To sort out the nature-nurture dilemma, ornithologist Amelia Laskey tried hand-raising a mockingbird in the late 1930s. One August morning she drove to a park five miles from her house, pulled a baby mockingbird out of a nest, and took him home to study. Honey Child was nine days old. (His observer was one of those scientists who could, as one writer put it, "stare unblinking at a bird's nest for days on end.") Like Jefferson's bird Dick, Honey Child would come to rule the roost until he died fifteen years later. His first tentative notes came when he was almost four weeks old. "He sang softly with closed beak for ten minutes," wrote Laskey, "a series of almost inarticulate warblings and whistles . . . utterly lacking in imitations of other species." Occasionally he ventured a very soft-toned "whisper" song of his own, a raspy cheep or chirrup, "an exquisite thing—soft, appealing, and infinitely tender in its cadences."

By four and a half months, Honey Child's songs were interspersed with the whistles, trills, warblings, and squawks of birds he could see or hear from his indoor home: downy woodpecker, Carolina wren, blue jay, northern cardinal, starling, bobwhite. During this first song season, household noises, particularly that of the vacuum cleaner, often started him singing. And as spring approached, his songs grew loud, varied, and long, beginning at 5:30 A.M. and continuing all day "like an aviary of chattering birds," said Laskey.

At nine months, Honey Child ventured his first direct imitation, responding instantly to a tufted titmouse for the first time with a *peto, peto, peto* song of his own. Eventually, he added to his repertoire dozens of birds (favoring especially the *wicka* of a flicker), along with the squeak of the washing machine downstairs and the whistles of the mailman and Mr. Laskey calling the dog. Some songs he would sing for a while, then drop them from his repertoire, only to resurrect them again the following spring. One June day, a count during sixteen minutes of lively singing yielded a list of 143 calls or songs of at least twenty-four other species, an average of nine per minute.

WE CALL THIS intricate process of vocal learning "advanced" or complex because it's done our way—through listening, imitating, and practicing. Lately, science has probed the deeper details of vocal learning in that obliging little bird from Australia, the zebra finch.

Dolphins and whales are good vocal learners, too, but for obvious reasons, they don't make very good lab animals. The ideal model organism for studying any kind of learning is a rare beast, says biologist Chip Quinn: It "should have no more than three genes, be able to play the cello or at least recite classical Greek, and learn these tasks with a nervous system containing only ten large, differently colored, and therefore easily recognizable neurons."

The zebra finch does not quite measure up, but it does make an excellent animal model for vocal learning. Named for the black and white stripes on their throats, zebra finches are easy to breed, mature rapidly, and are champion songsters, even in captivity. A young male zebra finch learns a single love song from his father or other males in his first ninety days after hatching and faithfully repeats that song throughout life. "Because it's impractical—and unethical—to monitor and manipulate the neurons implicated in vocal learning in humans," says Richard Mooney, a neuroscientist at Duke University, "these songbird tutors and students provide a beautiful substitute system so that we can study the detailed brain mechanisms that underlie this relatively complex type of learning"— from the stages in the process right down to the genes that flicker on and off while the bird is learning.

A BABY ZEBRA FINCH starts his long journey to the full-throated music of mature song just as we do on our journey to speech: He listens.

Incidentally, birds do have ears. Not our fleshy external pinna, just tiny holes beneath the feathers on either side of the head. The song a young bird hears sends sound waves into his ear and vibrates the hair cells there. These are ten times as dense as ours, and much more varied, allowing birds to detect high-pitched sounds beyond our range, as well as the soft rustling of insects beneath soil or leaves. (If a bird's hair cells are damaged by disease or loud noises—say, by the blasting decibels of a rock concert in a domed stadium—they can regenerate. Ours can't.) Sensory nerves in the brain stem pick up the signals from the hair cells and pass them along to auditory centers in the forebrain, where the neurons will form an auditory memory of the song.

In the first two weeks of its life, the baby bird sits in his nest, listening closely to a tutor, usually his father. He's silent, soaking up the sounds around him as a baby does. The father sings, the young bird listens and begins to memorize. He doesn't try to imitate the song yet; he just forms a mental template, or "picture," of it.

As he listens, his brain begins to bloom with networks of nerve cells. The networks grow into an elaborate constellation of seven discrete but interconnected regions highly specialized for producing song. This is his song system. In nestlings that have not yet begun singing, the regions are small. But over the next weeks and months, they grow in volume, cell number, and cell size.

In one region, the high vocal center (HVC), specialized cells make fine distinctions in the sounds the chick hears, noting even the slightest millisecond-length differences in the duration of song notes, and firing only when the notes fall within a narrow range. This is the same strategy of pattern recognition we humans use—called categorical perception—to spot subtle sound differences in language, say, between "ba" and "pa."

By the time the young bird makes his first effort at singing, a memory of his tutor's song is already present, solidified in small populations of highly selective neurons distributed throughout his song system.

A BABY ZEBRA FINCH in the wild grows up listening to songs of a smattering of different species, just as a mockingbird does. He's capable of learning any of them, yet he learns only the signature song of his species. Sounds from the world flood into the young bird's brain, but only those from his own species begin to carve out permanent traces. It's a perfect example of the intertwining of genes and experience.

When a young zebra finch first hears the song of his own species, his heart rate accelerates, as does his food begging. It's wired into him. As the songs he hears sculpt his growing brain, some channels—those tuned to songs of his own species—are preselected to become powerful rivers, the connections between the nerve cells in those pathways strongly reinforced, while the smaller tributaries, songs not part of his genetic heritage, quietly vanish.

This discovery—that some young birds are capable of learning almost any song they hear yet possess a genetic template that predisposes them to their species' song—has a human parallel. Young children have a remarkable capacity to acquire any of the world's six thousand human languages without formal training, which suggests that we're genetically predisposed for the task of language learning. Yet we learn only the language or languages to which we're exposed, underlining the importance of experience in the process.

If a bird has no tutor, it sings a song that's unrecognizable or only a poor rendition. Baby birds raised without any exposure to a tutor song sing abnormally, usually a very stunted, simplified version of the species song. This is true for humans, too. Children with normal hearing who are raised without any exposure to human speech utter abnormal vocalizations.

The window of song learning for a zebra finch is open only so long. When the young bird starts to sing, he'll imitate the tutor's song only from this early sensitive period. Then around the time he becomes an adult, the gates of song learning close. Why this is so is a puzzle that goes to the heart of our own learning—and its limitations.

One neuroscientist, Sarah London of the University of Chicago, has found a clue in zebra finches. "A tutor's song actually alters a young bird's brain in a way that affects his future ability to learn," she says. London's research has shown that young birds exposed to a tutor learn easily until they reach the age of sixty-five days. Thereafter, learning ability shuts down, and the bird's songs remain fixed for life. But young birds isolated from this song exposure can learn well even after sixty-five days. The experience of hearing another bird singing apparently alters the song-learning genes of the learning bird through "epigenetic" effects; in this case, says London, through the action of histones—proteins that coat DNA and allow genes to be turned on or off.

In birds like the mockingbird, canary, and cockatoo, the gates of learning stay open for longer, so they can continue to add new songs as they grow older. But learning is harder for adults than for juveniles.

We humans, too, are "open-ended learners." And like mockingbirds and canaries, for us the task of language learning grows more arduous as we age. Babies learn languages with incredible speed. In the first two or three years of life, they can, with little effort, become fluent in two or even three languages and forever after sound like a native speaker. After puberty, we have to work a lot harder to learn a foreign language and have difficulty speaking without an accent. Some of our neural circuitry becomes fixed during childhood—and for good reason. If our brains were constantly rewiring themselves, they would be neither stable nor efficient. We would learn everything but remember nothing. Still, wouldn't it be wonderful to be able to throw open those doors when we needed to—say, if we wanted to learn Urdu at age sixty? To my mind, a mockingbird's ability to sing thrush or chickadee at three or four years is not all that far off from a baby boomer taking on Cantonese.

IN THE SECOND PHASE of song learning, the young bird begins to explore his voice. At first, he whispers faint random, quavering notes like Honey Child's, amateurish murmurings or random screechy sounds like a young violinist testing his instrument. Meanwhile, the links between his higher brain regions and his motor regions strengthen, giving the young bird more and more control over his syrinx. Within a week or so, he brings the two sides of his syrinx into close coordination and begins to sing recognizable syllables, though in no particular order. He just takes all the sounds he has heard and memorized and tosses them out helter-skelter. These early efforts are known as subsong, and they're just like a baby's babbling—noisy, variable, exploratory. It's motor "play," which helps both birds and infants learn to control the muscles required for song and speech. Scientists have discovered that birds have a special part of their song-control brain circuit devoted to this subsong, separate from the part they will later use for practiced song. It goes by the tongue-twisting name of lateral magnocellular nucleus of the nidopallium, or LMAN.

The shift to genuine song takes place over the following weeks and months, as the young bird rehearses his tune tens of thousands, even hundreds of thousands, of times. Each time, he listens for errors and corrects them, matching his own vocalization with the song he has memorized. A song well sung offers its own reward, a "bolus" of feel-good chemicals such as dopamine and opioids. Dopamine may provide the drive to sing; opiates, the reward—the closer the match to the template, the bigger the reward.

Sleep seems to play a role in song learning for young birds, just as it does in human learning. A growing body of research suggests that the human brain continues to process the learning of a new motor skill after active training has stopped and during the sleep that follows it. This may be true for birds, too. Zebra finches practice their songs by day and sleep by night. After a young bird is exposed to a tutor's song, neurons in the song-production part of its brain fire in bursts of activity during sleep. The pattern of neuronal firing reflects the specific song learned, suggesting that it carries information about that song. After sleep, the quality of a young bird's song deteriorates but then improves with practice the following day. Oddly enough, the more severe the deterioration, the better the imitation of the tutor's song.

WHO IS LISTENING MAKES a difference in a young bird's performance. If he's singing alone, he's in tune-up mode. This is his undirected song. But if a female is around, he'll muster his best version and sing it over and over in directed song. Even if he's still at a stage when his singing is poor, he manages to direct his motor mechanisms to produce as perfect a song as possible.

"I've listened to the two versions of these songs, directed and undirected, for dozens of years," says Richard Mooney, "and I can't for the life of me tell the difference. But the females can. They care that males are performing in this more precise stereotyped way." Clearly, says Mooney, "there's a lot going on in a bird's song that human ears can't appreciate."

Brain-imaging studies by Erich Jarvis and his colleagues show that when a solitary male is singing undirected song to himself, the patterns of brain activity differ from when he is singing directed song to a potential mate. When a male bird sings solo, the brain pathways involved in vocal song learning and vocal self-monitoring light up, along with the vocal motor control pathways. (This is true, too, when he sings in the presence of another male.) But when he sings the very same song to a female, only the motor control pathways are active. These studies suggest an intriguing idea: that the mental and cognitive state of a male bird shifts when he knows he's being evaluated.

Mother finches also guide the learning of their sons by offering visual cues like strokes of the wing or fluffs of the feathers to shift a young bird's song pitch toward that of his father.

All of this is powerful proof that social cues shape learning behavior in birds—just as they do in humans. Babies don't respond so much to members of the opposite sex, but their babbling certainly improves in the presence of their mothers.

AFTER SOMETHING LIKE one or two million syllables of trial and error, the young bird sings a strikingly true version of his tutor's tune. The song "crystallizes" in a complex system of brain circuitry—but it's not static. For some songbirds, such as canaries, who learn new songs each mating season, the HVC cycles in size seasonally, growing in volume in the spring and shrinking in the later summer. At first scientists thought these changes arose only from the growth of new connections between cells. But then Fernando Nottebohm and others discovered that the bird brain actually adds new neurons to its song circuitry. "The recruitment of new HVC neurons is part of a process of constant replacement," says Nottebohm. By labeling these nerve cells with proteins that make them glow bright green, scientists can actually see this replacement happen in real time, watching the wandering neurons migrate into the HVC and form synapses with other neurons as a bird learns a new song. What the neurons are looking for and what determines where they end up is one of the puzzles being sorted out in the labs of the scientists assembled in the Georgetown auditorium. But we do know that this extraordinary kind of "neurogenesis" is probably common to all vertebrates, including humans.

DARWIN WAS RIGHT to call birdsong "the nearest analogy to language." Not only do birds and people learn song and speech through a similar process but both have "windows" of learning when the brain is most easily wired. In both, a parent or other tutor enhances the learning. While birdsong may not measure up to human syntax in complexity, elements of birdsong look something like it.

A new theory by Shigeru Miyagawa and his colleagues suggests that human language arose from a kind of fusing of the melodic components of birdsong and the more utilitarian, content-rich types of communication used by other primates. "It's this adventitious combination that triggered human language," suggests Miyagawa, a linguist at the Massachusetts Institute of Technology. In Miyagawa's view, human language has two layers: a "lexical" layer, where the core content of a sentence resides, akin to the waggle dance of honeybees or a primate's calls; and an "expression" layer that's more mutable and more closely resembles melodious birdsong. Miyagawa is not suggesting that birdsong literally gave rise to human language; the two systems of communication did not evolve from a common ancestor. But sometime in the past fifty thousand to eighty thousand years, he says, the two approaches to communication merged in the form of language as we recognize it today. "Yes, human language is unique," says Miyagawa, "but its two components have antecedents in the animal world. According to our hypothesis, they came together uniquely in human language." If this is true, the big question is *how* they came together, which remains a mystery. Still, I love this idea that the expressiveness of language may somehow incorporate or reflect the melody of birdsong.

There's more hard biological evidence to back up Darwin's assertion of the close kinship of birdsong and language: Both birds and humans use similar brain circuits to produce their vocalizations. Our brains have regions analogous to theirs: Wernicke's area, which controls our perception of speech, is like a bird's song-perception areas; Broca's area, which dictates our speech production, is like a bird's song-production area. But what's really similar in bird and human brains—and not found in species that are not vocal learners—is the presence of the song- or word-production areas and the connections, or pathways, linking song- (or word-) perception areas to these song- (or word-) producing motor areas.

In these pathways, millions of nerve cells connect and communicate so the brain can first hear sounds, then produce them.

"If the behaviors are similar and the brain pathways are similar," says Jarvis, "then the underlying genes may be, too." And indeed, that afternoon in Georgetown, Jarvis announces that a massive international effort to sequence the genomes of forty-eight species of birds has identified a set of more than fifty genes that flick on and off in the brains of both humans and songbirds in regions involved in imitating sounds, speaking, and singing. This difference in activity doesn't occur in birds that aren't vocal learners (such as doves and quail) or in primates that don't speak. So this may be a shared pattern of gene expression crucial to vocal learning in both birds and humans.

THIS NEWS BEGS the question of how the brains of humans and birds, species separated by so many eons of evolution, came up with such a similar solution for vocal learning. Why would we share similar genes and brain circuits?

Jarvis has a theory. In one recent imaging study conducted by his lab, he noticed that when birds hop, genes become active in seven brain areas that directly surround the seven song-learning regions of the brain. The brain areas involved in singing and learning to sing seem to be embedded in brain areas controlling movement. This suggests to Jarvis an intriguing notion, what he calls "a motor theory for the origin of vocal learning": Brain pathways used for vocal learning may have evolved out of those used for motor control. Many of the genes Jarvis found in that set of fifty that overlap between humans and birds are active in the same way: forming new links between neurons of the motor cortex and neurons that control the muscles that produce sound.

For Jarvis, who trained as a professional dancer, this is an exciting idea. "In a common ancestor of both birds and humans, there may have been a kind of ancient universal neural circuit that controlled limb and body movements," he proposes. In the course of evolution, this circuit duplicated and the new circuit was co-opted for vocal learning. (The concept of something new evolving out of something old, from existing building blocks, is a familiar one in evolution. Old structures shift and acquire new functions.) This duplication event took place at different times in birds and humans, Jarvis suggests, but the net result was the same: the ability to imitate sounds.

"It's a case of convergence," explains Johan Bolhuis, "of distantly related taxa coming up with similar solutions to similar problems."

In this way, vocal learning evolved at least twice and perhaps three separate times, once in hummingbirds and again, either in the common ancestor of parrots and songbirds or independently in both parrots and songbirds. In humans, the brain pathways used for gesturing may have been co-opted and used for speech.

"People have a hard time with this," Jarvis told me. "It's basically a humbling theory because it downplays the specialness of speech and of vocal learning circuits. But it's the best idea I can come up with to explain the existing data."

Interesting to note: Jarvis's lab also found that the vocal learning circuits of parrots are organized a little differently than those of other songbirds and hummingbirds. Parrots have a kind of supercharged "song system within a song system" that may help them pick up different dialects of parrot speech.

JARVIS'S MOTOR THEORY may explain *how* vocal learning evolved. But *why* is another matter. Why would nature favor as fancy a system as vocal learning in birds, of all creatures—along with all the complicated and expensive brain circuitry to support it? Why is it so rare? Jarvis has a theory about this, too.

IN SPRING, a male mockingbird seized by musical derring-do seeks a higher and higher perch until he sits on the topmost twig of the highest tree and, in Thoreau's words, lets rip "his rigmarole, his amateur Paganini performances." He even sings at night. He sings leaning forward, wings lifted slightly from his body, throat fully extended. It's as if his own singing turns him on. And perhaps it does. His glorious, frenetic, persistent song is a form of foreplay. It's a love song—and a dangerous one.

Up there on his open perch, exposed to the cruel eyes of aerial predators, he does nothing to blend in. On the contrary, he sings to stand out. If he were to repeat the same song over and over, he might have a chance of escaping the notice of a hunting hawk. But in creating one novel sound after another, he leaps to the foreground, as if to say, "Here I am! Here I am! Come get me! Come get me!"

This, says Jarvis, may be one reason vocal learning is rare. "All the varied vocalizations an animal learns make it an easy target."

Jarvis suspects that vocal learning may exist on a continuum among animals. "Some species—the advanced imitators like songbirds and humans—are at one extreme; and those with limited ability—including mice and maybe some other birds—at the other," he explains. Animals with complex vocal learning are usually either at the top of the food chain, such as humans, elephants, whales, and dolphins; or they're good at getting away from their predators, such as some songbirds, parrots, and hummingbirds. "Predators are actually picking off the others," he suggests. "To test that hypothesis, you would have to breed an animal for many generations without predators to see if vocal learning naturally evolves. That's a hard experiment to do but it's theoretically possible."

Research by Kazuo Okanoya of the University of Tokyo and his colleagues offers some evidence for this theory. Okanoya studies Bengalese finches, a domesticated strain of the wild-rumped munia, bred in Asia for their plumage, not their song. Okanoya found that Bengalese finches, kept in captivity for the past 250 years, sing more varied songs than their wild relatives. It's in part the relaxed pressure from predation, Okanoya suspects, that has allowed the domesticated birds to develop a bigger, more complicated repertoire. The females of both domesticated and wild finch varieties prefer the larger range of the domesticated song.

"So what I think is happening," says Jarvis, is that vocal learning is being selected *against* by predators—making it rare—but it's being selected *for* by sexual selection. Maybe that's how it worked in humans, too."

THE IDEA CAME to him one day while he was sitting in a park near the gardens at Duke University, reading. He heard a song sparrow singing from the top of a pine tree.

"I look up and I see him singing really loudly, boldly. He's singing the same song over and over again. I habituate to it. I continue to read. I don't pay attention. Suddenly the song switches. I look up again to see if it's a different bird, and I see that it's the same bird. Five minutes later, he's switching his song again, and I think it's another bird. He's keeping my attention. And I'm not even a song sparrow."

(This reminds me of a cartoon my ornithology teacher shared with our class. Two birds perch high up in the branches of a tree. Beneath them stand two birdwatchers with their binoculars pointed upward. One bird says to the other, "They still can't find us . . . let's sing something different!")

Singing is both risky and expensive. It not only makes a bird more conspicuous to predators but it takes time away from foraging. So why do birds bother?

Because songs well sung are the best tool out there for getting the girl, says Jarvis. "Vocal learning birds (and whales, too) change their vocalizations so they can be attractive to the opposite sex. Male birds sitting at the top of the tree in broad daylight, visible to the hawks and other predators that might pick them off, are saying to the female (to put it in anthropomorphic terms), 'Here I can sing boldly, loudly, and have all these different imitative sounds.' They're basically boasting: 'Look how well I can sing. Look how well I can imitate. Choose me.'" The mockingbird's Paganini performance with puffed-out chest is one big emphatic come-on, a "Hey, babe, check me out."

Extravagance in nature is so often found in proximity to sex.

For many male birds, competition for a mate is fierce. It pays for a female to be choosy. The stakes are high. She has a vested interest in selecting a male who's genetically fit and best able to defend her nest and foraging territory. One way she measures a potential suitor is by his song. If he's not singing the "right" stuff, she's off to find another date.

What is she listening for? (Or as Freud would ask, "What do women want?")

For a long time scientists assumed that the sheer size of a male bird's repertoire held cachet. But assessing how many songs a suitor sings is a difficult and time-consuming task. It's far easier to gauge how well he does from just a single song or two. Studies show that females of many songbird species prefer to mate with males that sing faster or longer or boast a more complex song. In other words, it's not how many songs he sings but how well he sings them.

Just what makes a song sexy seems to vary from species to species. Female swamp sparrows and domesticated canaries favor trill rates nearing the limit of possible performance, while zebra finches lust for loud songs. Some female songbirds have a soft spot for long or complex songs. Others, such as canaries, are turned on by "sexy" syllables. This is a real term in the field. A syllable is sexy when a male bird uses his syrinx to sing with two different voices at once. In a sense, it's like he's singing a duet with himself. Female canaries much prefer these sexy two-voice syllables to single-voice syllables.

Some females are sweet on the songs of the boy next door. They're looking for fidelity to local song or dialect.

Many songbirds have regional dialects with "accents" as distinct as a Boston "Southie" or an Arkansas drawl. These dialects are learned and passed along like family heirlooms from one generation to the next. A northern cardinal listening to recordings will respond much more vigorously to the voices of local cardinals than to those of cardinals from a habitat eighteen hundred miles away. The great tits of southern Germany have a dialect so distinct from the great tits of Afghanistan that the German birds don't recognize their Middle Eastern relatives. Even birds from different areas within a single state in the United States may sing entirely different tunes. According to ornithologist Donald Kroodsma, the black-capped chickadees living on Martha's Vineyard sing a different tune than their colleagues on the Massachusetts mainland. The geographical separation between song variations can be on the order of only a mile or even less. Among the white-crowned sparrows of California, for instance, distinct dialects may be separated by only a few yards. Birds that live on the cusp of two dialects are sometimes "bilingual."

Like the pronunciation, spelling, and vocabulary of human language, bird dialects may drift over time. Savannah sparrows, for instance, sing distinctly different songs today than their ancestors did thirty years ago. Some time ago, Robert Payne and his colleagues documented the cultural evolution in the songs of indigo buntings over a period of two decades. Each bird sang a song in the local tradition learned from his mentor, but with slight innovations. Payne used these markers to trace cultural descent of the bunting's songs. The vocal innovations, it turns out, persist in a population beyond the life of the bird that originated them. Eventually these create local song traditions and regional dialects that the birds recognize and distinguish.

And here's the point where females are concerned: Just as a Southie accent may not fly in Arkansas, songs that veer from the local dialect may be a buzz kill for female songbirds, possibly because a male who isn't singing the neighborhood song may have a harder time defending his territory.

IN JARVIS'S MIND, it all comes down to modulation. What makes a female swoon in the end is how well a male controls the tempo and precision of his notes, whether in a long, complex song or in a short, sexy syllable. "It's like a superstimulus," he says. "Like the allure of a big egg to a chicken." (As ethologist Niko Tinbergen learned, hens like big eggs:

Give a hen a giant egg to sit on, even an artificial one, and she will prefer it to a small egg. In her mind, bigger is better, even if it's not natural.) Some qualities are just irresistible. And for female songbirds, precision and meticulous modulation in song are about as sexy as it gets.

The precision of a bird's song is staggering. Richard Mooney makes the case to his colleagues at the Georgetown conference by showing two spectrograms side by side. The one on the left displays the vocal patterns of a human speaker asked to repeat a simple sentence a hundred times over; the one on the right, the patterns of a zebra finch from his lab singing his stereotypical sequence of syllables and motifs again and again. ("You have to pay a human to do this," Mooney quips; "the zebra finch does it for free.") This isn't just any human; he's a PhD candidate in neuroscience in the audience with us, and he's a straight-A student, "very, very articulate," says Mooney. "I asked this student to repeat as precisely as possible the sentence 'I flew a kite.'" (He picked the "I" sound, he says, because it has a pitch close to one of the zebra finch syllables.) "The zebra finch had no instruction at all."

Compare the two spectrograms side by side and the results are clear: No matter how hard the diligent student tries, his replications of his own syllables are wildly variable. The zebra finch's are nearly identical. In terms of his precision, says Mooney, "the bird is like a perfect machine."

It's what's known as vocal consistency, the ability to perfectly replicate the acoustic features of a song—the notes, the rhythms, the pauses—from one rendition to the next. To a bird, these subtleties make all the difference.

Consider what's involved in this sort of precision: the nervous system issuing exactly the same commands to the vocal motor system over and over again; accurate coordination of the muscles of the right and left sides of the syrinx, as well as those of the respiratory system, all in a matter of milliseconds; capped off with plenty of stamina so muscles don't fatigue. All in all, not a bad measure of a male's vocal prowess.

And indeed, females seem to use precision as a reliable gauge of male vocal performance. Lab studies show that zebra finch females strongly favor males with more consistent courtship songs. Male great reed warblers with more uniform whistle notes have bigger harems. Likewise, male banded wrens and chestnut-sided warblers with unswerving songs get more extra-pair copulations and sire more extra-pair offspring. The same holds true for mockingbirds—those with more consistent songs father more young than, and are dominant over, males with sloppier songs.

SCIENTISTS ARE still sorting out just what all this precision and fidelity actually signal to a choosy female songbird. Superior song performance may be a signal cuing her that a male is physically fit. A strong, unwavering song with superior amplitude, duration, and consistency may be a male songbird's way of saying he has good motor control and his body is in fine physical condition. A bird of lesser mettle couldn't muster such a performance. Other qualities, the so-called structural traits of his song—how accurately he sings his tutor's songs, whether the syntax of his song makes sense, and how complex it is—may be his way of saying he was well fed as a nestling and free from stress (or able to withstand it) and, as a result, has good brain structure and functioning. Sexy syllables in canaries, for instance, require extraordinary coordination of the left and right halves of the syrinx. Listening for supersexy syllables allows female canaries to rule out males with poor bilateral coordination.

Because birdsong is such intricate and demanding behavior, it may be a handy and sensitive barometer not only of a suitor's overall health but also of his brainpower.

It goes back to those critical windows when a baby bird is furiously generating the connections to form the song systems in his brain, says Steve Nowicki of Duke University. At the same time, his body is also growing like mad. A typical songbird nestling reaches about 90 percent of his adult weight within the first ten days of his life—an incredibly rapid rate of growth. All those neurons, muscles, feathers, and skin require plenty of nutrition. So it's a vulnerable time. If something happens during those precious weeks—if parents can't deliver sufficient food or if the young songbird endures disease or other kinds of stress, such as competition from siblings—the song circuits in his brain suffer. Birds in captivity that are underfed develop atrophied song structures in the brain and don't copy their tutor's song as well. One study, for example, showed that well-fed zebra finches copied 95 percent of syllable types from their tutors while underfed birds copied only 70 percent. It may not sound like a big deal, but to the females, it matters. She can "sniff" out missteps in his song and judges him harshly for it. In other words, a male bird is deemed only as good as his song. His melody betrays his biography for his entire life.

A sparkling song precisely sung, then, may signal a male's superior brainpower and capacity to learn. This "cognitive capacity hypothesis" suggests that a female chooses her mate based on smarts, using his song as a proxy. In other words, birds that sing better are showing females they're good learners. A superior singer is not only better at acquiring, memorizing, and faithfully producing fancy songs, the theory goes; he's also likely better at other brain-powered tasks—all kinds of learning, decision making, and problem solving, such as where, when, and what to eat, how to avoid predators, and how to attract mates—presumably premium traits for a female desiring "good" genes and/or a proficient food provider for her offspring. However, it's not clear whether a male's song performance actually correlates with his performance on other cognitive tasks. The evidence is mixed.

When Neeltje Boogert of the University of St Andrews studied isolated male zebra finches in the lab, she challenged them with a single task: prying the plastic lid off a wooden well to get at food. She found that birds that sang more complex songs, containing more elements per phrase, were quicker to figure out the task than males who sang songs with fewer elements. This suggests that females may be using a male's songs to judge his foraging smarts—how well he learns where and how to find food.

But the story is not so simple. When Boogert and her colleagues later tested male song sparrows (which sing a larger variety of songs than zebra finches) on a greater range of cognitive tasks—such as reversal learning and spatial and color association tasks—the more expert singers got mixed reviews. They did better on some tests and worse on others. And recently, in a study of zebra finches in a flock—their more natural, social context—the correlation between song complexity and other cognitive abilities vanished. The finest songsters did no better on problem-solving tasks than the more mediocre performers. Confounding factors may muddy the picture, says Boogert, variables such as stress, motivation, distraction, and social dominance.

It's perhaps even trickier to test for possible correlations between song performance and cognition in the wild. Not long ago, Carlos Botero took an unusual approach to the problem. The intrepid researcher, then at the National Evolutionary Synthesis Center in North Carolina, trooped through desert, jungle, and scrublands in several South American countries carrying sensitive recording equipment to capture the songs of mockingbirds in the wild. After recording one hundred tracks from twenty-nine species, he found that mockingbirds living in unpredictable climates sang more elaborate songs. In fickle environments where capricious weather—erratic rainfall and fluctuating temperatures—made food sources iffy, mockingbirds not only had a bigger repertoire but were better at copying the songs and calls of other species, with truer notes and more consistency. Perhaps a male's singing skills signal to females that he's clever enough to cope with unpredictable environments, says Botero. This adds weight to the idea that some aspects of birdsong may provide information about a male bird's general cognitive skill—and that sexual selection is acting on these signals of intelligence.

IT'S LATE AFTERNOON, hours after that first break in the birdsong meeting. I duck outside again to check on the cedar tree. The mockingbird is still in his sheltered perch, singing his myriad tunes, but this time pianissimo.

Whether female songbirds use male song as a proxy for general smarts remains to be determined. But one thing seems clear: Over evolutionary time, females have shaped the complex, precise, extravagantly beautiful tunes of their kind—and the elaborate brain circuitry required to produce it. As ornithologist Donald Kroodsma explains it, in listening to a male's song and evaluating it, the female "designs" him so that his singing tells her whether he is worthy to be the father of her offspring: "By the mating choices she makes, she perpetuates the genes of 'good singers,' with 'good' being defined by something that lies deep in the female psyche of each species." In this sense, then, the female sculpts in a male a neural song network of miraculous complexity and a brain that rewards him for the precision of his own song. It's known as the mating-mind hypothesis: Cognition for complex male displays and cognition for female evaluation of those displays evolve together, affecting the brain structure in both sexes.

There's no female in sight of this crooning male in the cedar tree. Maybe his fall song offers that other kind of prize. Birds that sing their songs well in spring and fall experience those rewarding chemicals, dopamine and opioids—but in different amounts in each season, and to different ends. Opioids induce not only a feeling of pleasure but also analgesia, says Lauren Riters. To find out which season's song produced more painkilling opioids, Riters observed male starlings singing in fall and spring, captured them, and then dipped their feet in hot water. She predicted that birds singing fall song would endure the heat longer. She was right. Fall song, she found, is more tightly coupled to opioid release than spring song. As Darwin wrote, "the songs of birds serve mainly as an attraction during the season of love," but after the season for courtship is over, "male birds . . . continue singing for their own amusement." Or possibly for the drugs.

This bird in a bush is not in full tenor mode. Though his song is still filigreed with imitation, it's sung with such quiet grace it seems he must be singing to and for himself. Perhaps to numb the cold. That's plausible. Or perhaps because when he sings his sweet trilling notes precisely, beautifully, it both eases his pain and quite literally packs him with pleasure.

six

THE BIRD ARTIST

Aesthetic Aptitude

C rouch behind the buttresses of a blue quandong tree and watch through a weft of branches. On a sun-mottled spot of the rain-forest floor is a bird about the size of a pigeon but glossy blue-black with a bright purple eye. Behind him is an elegant little twiggy architectural hall about a foot high, formed by two parallel arched walls of upright sticks, like a toy teepee built by a child. All around him, the ground is stippled with colorful objects that pop out against a carpet of buff-colored twigs, almost glowing in the dim light of the forest. There are blossoms, fruits, berries, feathers, bottle caps, straws, the wings of a parrot, a tiny toy Bart Simpson skateboard, and what looks for all the world like a turquoise glass eyeball. The bird picks up a flower and drops it nearby. He arranges a feather, nudges a bead, pokes at a straw—apparently sorting his booty by color, size, and shape. Every so often he hops back, as if to survey his handiwork, then hops forward again to rearrange a piece.

If you had been watching this bird a few weeks earlier, here on the east coast of Australia, you would have seen him in a pitch of industry. First, he furiously clears debris from an area about a yard square and then sets about diligently collecting twigs and grasses, which he distributes evenly to make his "platform." From this collection, he selects choice twigs to plant in two neat rows, creating a kind of avenue carefully positioned to catch the morning strike of sun. At the northern end, he arranges his bed of fine twigs, evening it out. This will serve as the background for his panoply of decorations—and also as a kind of dance floor, where he will later offer up some showy pirouette and song.

Next comes the business of collecting treasures. Not just any object will do. This bird is bullish on blue: cornflower-blue tail feathers from a parrot, lavender lobelia blossoms, shiny blue fruits from the quandong tree, purple petunias, and blue delphiniums stolen from a nearby homestead, along with fragments of cobalt glass or pottery, navy blue hair ribbons, bits of turquoise tarp, blue bus tickets, straws, toys, ballpoint pens, that eyeball, and his prize, a baby-blue pacifier pilfered from his neighbor. These he arranges artfully against his twig canvas. If his flowers wither or his berries shrivel, he'll replace them with fresh ones. Watch for a few more days, and you might see him paint a chest-high band on the inside of his twig hall, using dried hoop pine needles he has chewed and crushed in his beak.

No wonder early European naturalists were perplexed when they found these creations deep in the Australian forest and thought they had stumbled on fanciful dollhouses made by aboriginal children or their mothers.

WE ARE AWED BY animal builders, perhaps because we're builders ourselves. It's why we marvel at that most familiar piece of avian architecture, the bird nest—especially the ornate constructions of certain species: weaverbirds, for instance, who entwine and knot plants to build their elaborate nests; or Baltimore orioles, who knit together their nests with tens of thousands of rapid shuttle stitches; or barn swallows, who make thousands of trips with mouthfuls of mud to fashion their cup-shaped nests on the rafters of barns or the underbellies of wharves and bridges.

"The implement which determines the circular form of the nest is no other than the bird's body," writes Jules Michelet. "His house is his very person, his form . . . and I would say, his suffering."

I thought of this when I saw the tiny cup nest of a white-throated fantail perched atop a single stalk of pandanus along the banks of a river in the Tanjung Puting region of Borneo. This fantail is a common bird of the open forest there, but its compact little nest is a marvel of clever construction and delicate engineering, perfectly round, barely large enough to hold the mother and her baby bird. I wondered if the birds had pressured the walls with their own breasts and used their bodies to press and knead the materials until they were pliable. The nest was anchored on the top of the stalk with lines of spider silk and bracts of coarse grass, its walls plaited with fine grasses, tiny overlapping leaves, the down of tree ferns, and threadlike roots to form a snug, spherical cup.

An award for nest-building brilliance should go to the long-tailed tit, a common relative of the chickadee that lives in Europe and Asia. Its nest is a flexible bag composed of small-leaved mosses that form hooks, which are woven together with the silk loops of fluffy spider egg cocoons to create a kind of "Velcro." The little birds work to line the inside of the bag with thousands of small insulating feathers and cover the outside with thousands of small lichen flakes for camouflage, creating a structure made of roughly six thousand separate pieces.

"A bird's nest is the most graphic mirror of a bird's mind. It is the most palpable example of those reasoning, thinking qualities with which these creatures are unquestionably very highly endowed."

The English ornithologist Charles Dixon wrote that in 1902. Nevertheless, we've long assumed that nest building is a purely innate behavior: A bird comes into the world with a kind of nest "template" wired into its genes and no real concept of what it's aiming to make. If brains were involved at all, it was only in adhering to a simple set of rules for how to behave, programmed movements that led to the emergence of the fancy egg cup. Nobel laureate Niko Tinbergen noted that long-tailed tits use a sequence of up to fourteen motor actions to construct their domed nests, but then remarked how amazed he was that such "simple and rigid" movements together "could lead to the construction of so superb a result."

Lately this view has changed, as scientists have stockpiled convincing evidence that nest building requires all sorts of qualities other than instinct—for instance, learning and memory, experience, decision making, coordination, and collaboration. The magnificent creation of the long-tailed tit, it turns out, is a cooperative effort between the members of a pair from start to finish. It's work that requires a suite of decisions about location, materials, and construction itself.

It makes sense, then, that when Sue Healy, a psychologist and biologist at St Andrews University in Scotland, and her Learning and Nest Building team probed the brain regions a zebra finch uses during nesting behavior, they found activity not just in the motor pathways of the brain but also in pathways involved in social behavior and reward.

In an experiment reported in 2014, Healy and her team asked the zebra finch if it could learn to choose more effective nesting material based on experience. In the wild, zebra finches build their nests in dense shrubs from hollow balls of stiff dry grass stems or fine twigs. In the lab, the scientists gave the birds nesting materials of either flimsy, flexible cotton string or a much stiffer variety. After a brief bout of nest building, the birds were given a choice of the two strings. Those that had experience with the flimsy string eschewed it in favor of the more rigid variety. The more nest-building experience the birds gained, the more they opted for the stiffer string. Clearly, learning influenced their choice of building supplies.

To see whether the birds deliberately select materials that camouflage their nests, the team "wallpapered" the cage of male zebra finches in different colors. Then they gave them a choice of nest-building materials: strips of paper that matched the wallpaper and strips of a different color. Most of the birds picked the matching strips, suggesting that they scrutinize the characteristics of their building materials, and don't just serendipitously pick up whatever may be available locally.

Village weaverbirds, too, learn to improve their choices of nesting material with experience. Young birds prefer to build with more flexible materials and longer rather than shorter strands. As they gain experience, however, they grow pickier, rejecting anything artificial, like string or raffia or toothpicks. They get better at cutting and weaving as well, making fewer mistakes and creating neater, more tightly woven nests as they get older.

THE TWIG-AND-OBJECT CONSTRUCTION of that glossy bird Down Under, however, is not a nest. That bird, unlike the cooperative long-tailed tit, leaves nest building entirely to his mate. No, this strange and elaborate creation, known as a bower, is built for one purpose only— seduction—by a creature of extraordinary craft and intelligence, the satin bowerbird (*Ptilonorhynchus violaceus*).

So remarkable is the bowerbird family that the ornithologist E. Thomas Gilliard once remarked that birds should be split into two groups: bowerbirds and all other birds. Bowerbirds are noted for the hallmarks of intelligence: large brains, long lives, and extended periods of development. (It takes them seven years to mature.) All twenty or so species live in the rainforests and woodlands of New Guinea and Australia; seventeen species build bowers. They are the only animals on the planet—except us, perhaps—known to use objects in extravagant displays to lure mates.

AND THERE SHE IS. A dull olive-green bird about the size of her suitor. She has been touring the neighborhood, scoping out the workmanship of three or four other bowers and assessing their decorations.

It's her market, so she's shopping around. She lands just south of our hero's bower and hesitates in the undergrowth. She seems to like what she sees. Maybe it's the satisfying symmetry of his architecture that catches her eye. Or that powder-blue pacifier. A few moments later, she hops into the cozy little crafted bower and nips at a few sticks, tasting the paint he has carefully applied to the interior of the bower walls.

As soon as she lands, the male halts his tidying and quickens. He leaps into a frenetic ballet of hops and dances. With his beak, he plucks up objects from his prized collection and drops them around the floor of his stage. Suddenly, he goes "mechanical," buzzing and whirring like a rhythmic wind-up toy. It's less croon and swagger than herky-jerky robot or mannequin. He flicks his wings and fans his tail in quick, shutterlike movements, then runs dramatically across his platform as if attacking an aggressor. Abruptly, he launches into a torrent of mimicry. First, the rolling cachinnating call of a kookaburra, then the rattling machine gun of a Lewin's honeyeater, then the softer calls of a sulphur-crested cockatoo, an Australian raven, a yellow-tailed black cockatoo. He chortles. He buzzes. He squeaks and *chur*s. He flaunts his splendid plumage and flashes his bulging eyes, now strangely suffused with red. He pauses, staring fixedly, hops around for a few minutes, then abruptly resumes his display. He thrusts his neck forward and flicks his wings again. Snatching up a small decoration in his beak—a yellow leaf—he hops stiffly to the bower entrance and faces the female, puffing up his glistening feathers so that he looks bigger, and performing a sequence of deep knee bends.

The female watches all this showmanship intently, gauging his performance, which may last a few seconds or as long as half an hour.

Suddenly our hero shies violently sideways. The female startles. In an instant, she flits out of the bower and away.

He has lost her.

Why? Where did he go wrong?

THE HARD TRUTH in the universe of bowerbirds is that relatively few guys get the girl. It is the females who exercise choice in their amours, and they make a very careful selection. Often, one male is lucky many times over, mating with twenty or thirty different females, while other males get no matings at all. The reasons for this inequity are complex and offer an intriguing window into how the bowerbird developed his artistry and smarts. How did a male's penchant for dance and finicky displays of symmetrical twigs and cerulean straws get cross-wired with a female's notion of an ideal mate? Is his "artistry" an indicator of intelligence or aesthetic sense?

The story of the satin bowerbird is a good place to search for answers to such questions. These bowerbirds exhibit pretty extreme display behaviors, says Gerald Borgia, a biologist at the University of Maryland who has studied the birds for more than four decades. And females display pretty extreme choosiness.

What are they looking for?

Male bowerbirds don't offer any direct benefits as helpmates. No assistance with feeding young, for instance, or protecting territory. The only thing a female bowerbird gets from a male is his genes. So she doesn't waste her time assessing his ability to, say, forage. Instead, she scrutinizes his bower and decor, as well as his skill at dancing, mimicking, and other courtship displays.

Shopping around takes time and energy, so these demonstrations must signify. Indeed, in every aspect of his display, says Borgia, a male reveals his nimbleness of mind.

Consider what it takes to build a superior bower:

First, the choice of an excellent location. A savvy male positions his bower to maximize the appeal of his display. The satin bowerbirds that Borgia studies orient their bowers on a north-south axis. "They seem to be trying to get the illumination just right for their displays," says Borgia. Sometimes they prune the leaves around the platform to let in more sunlight.

Second, fine craftsmanship. Females prefer bower walls that are superbly crafted, symmetrical, and dense with uniform sticks. So a hopeful male must find straight, slender sticks of just the right length, hundreds of them. These he packs tightly into two thick, curved walls. To make the walls symmetrical, he uses a mental tool called templating. "In templating, a male picks up a stick and positions himself along the midline of the bower avenue," explains Borgia. He puts the stick in or against one wall and, still holding on to it, pulls it away from the wall—then, using a precise reversal of his movements, he places the stick in an identical position in the opposite wall. Some bowerbirds are flexible enough to modify the technique. When experimenters messed with the bowers of several males, completely destroying one of the symmetrical walls, the birds revealed a surprisingly agile mind: Instead of placing half the sticks evenly on each side, they concentrated their efforts on rebuilding the destroyed single wall.

Then there's the deep matter of adornments. Females like decorations, a lot of them, so males hoard their bling. Remove these treasures from a male's bower, and his stock drops precipitously. He constantly adds to his collection—sometimes in unscrupulous ways—stealing from neighboring bowers if their owners happen to be away. Keeping his own bower intact and smartly appointed absorbs all of his energies.

Each species of bowerbird has its preferred ornaments and colors carefully selected for contrast, depending on the environment. Spotted bowerbirds, the scrappy cousins of satin bowerbirds that build their bowers in open woodlands, favor green objects, along with silvery, shiny things, says Borgia. "They put coins, jewelry, and new nails in the sweet spot of their bower and rifle cartridges further out. We found one bird who had put shiny new nails in the bower avenue and oxidized ones in the back. He was sorting the nice from the not-so-nice ones." These birds often locate their bowers near "tips," the Australian term for dumps, where they have ready access to all kinds of glinting, colorful stuff. One bower Borgia discovered, built by a spotted bowerbird near the house of a stained-glass artist, was filled with little shards of stained glass, which the bird had sorted by color. "It was remarkable how he laid out the pieces," says Borgia, "just like a mosaic."

The Vogelkop bowerbird builds a tall wigwamlike structure known as a maypole bower around a sapling tree trunk in the mountain rainforests of New Guinea. The roof is woven from the stems of epiphytic orchids. On a lawn of moss extending from the bower, the bird creates a beautiful still life of little piles of brightly colored blossoms, fruit, and iridescent beetle wings—red, blue, black, and orange—along with the occasional random treasure prominently positioned, such as a white tube sock with orange stripes stolen from a nearby missionary's cabin.

Great bowerbirds, which live in the eucalyptus woodlands of northern Australia, favor a minimalist raiment for their background, largely white stones, bones, bleached snail shells. (During a heavy storm at her Queensland field site in December 2014, Brazilian researcher Aída Rodrigues noted that the great bowerbirds there were even incorporating giant hailstones into their display courts.) The pale canvas heightens the contrast for the shiny objects they place at the avenue entrance, the green ones they carefully arrange in lines or ovals on the sides of the avenue, and the red ones scattered at the court edges.

These bowerbirds build two elliptical courts linked by a long avenue of brownish-reddish sticks, packing their bowers with an astonishing five thousand twigs. The female stands in the middle of the avenue while the male courts her. The reddish light from the sticks in the avenue may actually alter her perception of color, heightening her experience of red, green, and lilac—the color of the male great bowerbird's nuchal crest. He stays just out of sight in one of the courts where his colorful objects are stashed. At intervals, he pops his head around the corner to surprise her by tossing an object her way. It's his way of holding her attention. The longer she remains in the avenue, the more likely she'll mate with him.

According to John Endler of Deakin University in Australia, great bowerbirds may have another artistic trick up their sleeves: optical illusion. To impress the ladies, Endler argues, the males arrange their collections of stones and bones in increasing size with distance from the avenue entrance. In Endler's view, this sets up the perfect conditions for a visual illusion known as forced perspective.

It's a ruse similar to the one used by ancient Greek architects in their design of building columns tapered at the top to create the impression of greater height, and more recently by the designers of Disneyland's iconic Cinderella Palace. The bricks, spires, and windows of the blue and pink castle get smaller with each consecutive story, so your brain is tricked into thinking that the top of the building is farther away than it is. Filmmakers used the trick in *The Lord of the Rings*, too, to make the hobbits look smaller.

Great bowerbirds apparently do just the opposite: They put smaller objects closer to the bower entrance and bigger stones and bones farther away. To the female looking out from her cozy enclosure, the researchers speculate, this creates the illusion that the court is smaller than it is. The foreshortened stage may make the parading male himself and his colored objects look bigger and more vibrant. The female's brain, like ours, may make false assumptions about what she is seeing. To be sure of this, however, more research on bird perception is needed.

What kind of brainpower might be required by the male to pull off this bit of visual deception—if that's what it is? It could be a simple matter of trial and error, says Endler, with the birds putting objects at random and then going in to have a look and rearranging them. Or they could be using a simple rule of thumb—put smaller things closer, bigger things farther away—a slightly more complex behavior. Or maybe they have a true sense of perspective and know the order they should put objects in to construct a gradient. One thing we know for sure, says Endler, "the arrangement doesn't happen by chance." The birds are deeply committed to their own designs, Endler has found. When he and his team rearranged the white and gray objects in the bowers, the males put things back according to their original design within three days.

SATIN BOWERBIRDS ARE primarily colorists and choose their colors for maximum contrast. In crafting their bower stage, they lay down a carpet of light-colored twigs and leaves that generates a bright glow in the dusky ambience of the forest. Against this backdrop, they garnish their stage with blue, the rarest of nature's colors. Some scientists suggest that the satin's color scheme might be aimed at matching its own iridescent raiment. But Borgia has found that the birds have no interest in decorating their bowers with their own feathers. They just favor blue, which contrasts so well with buff in the green gloom of a rainforest.

Humans seem to love the hue, too. Surveys suggest that blue is beloved by more people than any other single color, perhaps because it's associated with beloved objects in the environment, clear skies and clean water. The painter and colorist Raoul Dufy reportedly said that "blue is the only color which maintains its own character in all its tones . . . it will always stay blue." In nature blue is unusual in part because vertebrates never evolved the ability to make or use blue pigments. The deep electric blue an eastern bluebird carries on its back is an example of what scientists call a structural color: It's generated by light interacting with the three-dimensional arrangement of keratin in the bird's feathers.

Blue objects are relatively rare in a satin bowerbird's world, so the birds often procure them by theft. The store of blue decorations in a male's bower reflects his ability to nip them from the large caches of objects in the bowers around him. Once acquired, these treasures must be guarded against other male bowerbirds keen on taking them for their own stash.

Some males visit rival bowers not only to steal but to destroy. This, too, requires quick-wittedness. The bowers of satin bowerbirds are generally more than three hundred feet apart, out of sight of one another. According to Borgia, ransacking a neighboring bower that is not within obvious view suggests that males possess a mental map of bower locations and remember them.

Borgia's research team uses surveillance video cameras to catch the vandals in action. A marauder seeks out and targets another male's bower with stealth and speed. He flies in silently and perches motionless in the branches above the bower, making sure the owner is away. Then he drops to the ground. In an instant he becomes a dark velvet tornado of activity, plucking sticks from the bower and flinging them aside. In a matter of three or four minutes, an architectural triumph that took days to construct is flattened to a heap of twigs. The brigand steps back to survey the wreckage, spots a blue toothbrush to loot, and away he goes.

From a female's perspective, an intact bower embellished with abundant blue objects suggests that this male is adept not only at thievery but at protection against pilfering and vandalism.

If satin bowerbirds favor blue, they reject red. Plop a crimson object down amid the blue ones, and the birds will quickly remove the offender, fly off with it, and drop it at some distance, out of view. Some observers even go so far as to suggest that fouling the bird's bower with any shred of red will make the bird madder than a wet hen.

Why this aversion to red? Borgia thinks the satin bowerbird's color combination of blue against yellow—not otherwise found in the bird's habitat—provides a clear, distinctive signal, a kind of flag for visiting females that calls out, "Here's a bower of your species!" Anything red is a polluting presence that disrupts the clarity of the signal.

The satin bowerbirds' urge to rid their bowers of red gave Jason Keagy, then a doctoral student working with Borgia (now at Michigan State University), an ingenious idea: Use this aversion as a powerful motivator to test the problem-solving ability of different males in the wild.

Keagy wanted to find out whether some males were smarter, and if these same birds won the most matings.

In one test, he placed three red objects in a satin bowerbird's bower and covered them with a transparent plastic container. Then he measured how long it took the bird to remove the barrier so he could get rid of the red objects. Some birds took less than twenty seconds to solve the puzzle; others couldn't do it at all. Most of the birds that solved the puzzle pecked at the container until it fell over and then spirited away the red objects. But one bird perched atop the container and rocked it until it toppled, then dragged the container away from his platform before disposing of the aberrant red.

The second test was a bit more devious. Keagy glued a red tile to long screws, which he screwed deep into the ground to make the tile immovable. This presented the birds with a novel problem, one they would not ordinarily encounter in their natural environment. The cleverer males quickly discovered a novel strategy to deal with the situation—cover up the red with leaf litter or other decorations.

Then Keagy correlated ingenuity in the two tasks with mating success. The speediest problem solvers in both tasks, it turned out, were also the mating champions, scoring many more copulations than the less competent birds. In other words, says Keagy, "Smart is sexy!"

IS A BOWERBIRD'S BOWER ART? Is the male bird an artist?

It depends on how you define *art*. Like intelligence, the term eludes easy definition. The *Oxford English Dictionary* defines it as a "skill, especially human skill as opposed to nature; cunning; imitative or imaginative skill applied to design." *Merriam-Webster* says it's "skill acquired by experience, study, or observation," and "the conscious use of skill and creative imagination."

Biologists offer a different perspective. John Endler suggests that visual art can be defined as "the creation of an external visual pattern by one individual in order to influence the behavior of others, and . . . artistic skill is the ability to create art." Richard Prum, an ornithologist at Yale University, views it as "a form of communication that coevolves with its evaluation." By these definitions, a bower would certainly seem to qualify as art, and bowerbirds, as artists.

There may be artistry in other bird creations. Some birds decorate their nests exuberantly. Black kites favor white plastic; owls, feces and remnants of prey. Many birds have an eye for glittery, shiny things. In his *Birds of Massachusetts*, Edward Forbush noted a report of one male Baltimore oriole who spied a child playing with a silver shoe buckle tied to a ribbon. The bird swooped down and seized the buckle, then wove it into his nest. On the Delaware shore, I've watched osprey bring to their nests shiny ribbons, bottles, and bits of Mylar balloons. One osprey nest in Monmouth Beach, New Jersey, had a wristwatch hanging from it.

Other birds may or may not be drawn to glittery treasures for aesthetic reasons. Only bowerbirds lavishly decorate display areas, seeking out treasures of specific colors and positioning them meticulously to bedazzle females. The naturalist and filmmaker Heinz Sielmann once observed the decorating behavior of the yellow-breasted bowerbird: "Every time the bird returns from one of his collecting forays, he studies the over-all color effect. . . . He picks up a flower in his beak, places it into the mosaic, and retreats to an optimum viewing distance. He behaves exactly like a painter critically reviewing his own canvas. He paints with flowers; that is the only way I can put it." According to Gerald Borgia and Jason Keagy, a male satin bowerbird does something similar: He sits in his bower where the female will sit, as if he were attending to her viewpoint, and then changes the display accordingly. "We're not saying these birds have theory of mind," says Keagy, "but it is a very interesting behavior nonetheless."

What would you call this collecting and sorting and carefully considered placing of colored objects with no apparent function other than to impress a viewer or evaluator and modify her behavior? To my mind, it seems almost impossible to perceive it except in some relation to art.

SO WHERE DID OUR spurned male hero go astray? His bower was a paragon of symmetry and artistry. His bright stage was stippled with blue come-ons looted from his rivals. He demonstrated remarkable mastery of vocal mimicry and dance.

But the lady of the satin bower, it turns out, wants something more.

Gail Patricelli, an animal behaviorist at the University of California, Davis, suggests that successful courtship among satin bowerbirds is not just a matter of brute smarts, artistry, and bravado. Something else plays prominently into the picture—something like sensitivity.

Females are drawn in by displays of song and dance that are vigorous and intense, but not excessively so. Immoderate wing flipping and buzzing can look a lot like an aggressive display of one male toward another, which is a powerful turnoff for a female. So males are in a bit of a bind, says Patricelli: They need to display intensely to be attractive, but not go over the top, or they may drive females away. Courtship calls for more sensitivity than swagger, more tango, less kickboxing.

To see how different males handle this dilemma, Patricelli created an inspired experiment while she was a doctoral student in Borgia's lab. She built a little "fembot," a robot tucked into the skin of a female bowerbird. She outfitted the birdbot with several small motors so that she could make it crouch like a real female bird, look around, and even fluff its wings in the mating position. This was Patricelli's way of controlling the variable of a female's behavior so she could measure male responses. The fembot did its thing the same way each time, and Patricelli video-recorded the responses of twenty-three different males.

The videos revealed that males vary a lot in their sensitivity to how a female responds to their display. Some males are attentive. If a female seems alarmed, they will rein in their display, tempering their wing flipping and giving her some distance. Other males are oblivious.

The responsive dudes, it turned out, are those that secure the most matings. Males who go overboard in demonstrating their intensity and power lose out. In other words, says Patricelli, sexual selection seems to favor both the evolution of elaborate display traits and also the ability to use them appropriately. And this may be where our hero fell short. He lacked social grace.

BUILDING, DECORATING, fine-tuning your song and dance display and moderating its intensity to suit your date: According to Gerald Borgia, these are extravagant behaviors a satin bowerbird is not born with but must acquire when he's young. Here, perhaps, is another clue for females: The quality of a male's display, like the precision of a songbird's song, indicates his ability to learn as a juvenile. And, as with song learning, this may be a sign of his cognitive capacity.

The genetic payoff for being one of the lucky guys is big. Very big. So males work hard at learning to build the finest possible display and intensely practicing their courtship skills. In fact, these birds devote little of their waking time to anything else.

"Young males build rotten bowers," says Borgia. Because they fail to select sticks of different lengths and sizes and place them at proper angles to produce rounded walls, as adults do, their bowers end up a sloppy mess. "Also, the juveniles will use ridiculously thick sticks," says Jason Keagy, making it even harder for them to build a nice, tidy bower. "And another funny thing," adds Keagy: "Juveniles will work together on the same 'practice' bower—but not collaboratively. So a male will add sticks; another male will come and destroy what's built and start over; then another will pop in and add a few more sticks, and so on."

The young birds get better at the task over time, mainly by emulating their seniors. They visit the bowers of other males and sometimes "help out" by building on to an existing bower or just adding a stick or two to the walls. They also paint on bowers belonging to other males. (Bower painting is an important part of the bower seduction. When experimenters removed this paint from the bowers of some males, fewer females returned to those bowers for second courtships and copulations.)

The juveniles watch their elders for tips on displaying, too. This requires a bit of fancy role-playing. When a young male visits the bower of a mature male, he often plays the female's part while he closely observes the older male. He may be a bit more fidgety than his feminine counterpart, but the older bird tolerates his presence because the mentor, too, benefits from practicing with a live audience. "It's a win-win situation," says Borgia; "otherwise you can bet it wouldn't happen."

Think of it. To win a mate, a male satin bowerbird must be artistic, smart, sensitive, athletic, handy, and a good learner. A choosy female, for her part, must have considerable brainpower to size up all these qualities. As Jason Keagy observes, mate choice is a demanding cognitive process. It involves narrowing down the candidates throughout the season, visiting male bowers for first courtships, and returning for more courtships before finally deciding to copulate with a particular male. A female must zero in on the location of male bowers, which are often well concealed beneath shrubs and sometimes a few miles apart—requiring a mental map of sorts—and she must remember them from season to season. She must assess building skills and count decorations, or at least estimate the numbers. She must taste the paint on that chest-high band lining the inside of the bower, possibly a chemosensory signal for her to use in assessing a male as a potential mate. She must evaluate a male's display, listening for accurate mimicry and gauging the vigor and skill shown in his fancy footwork and the intensity and power of his performance— all the while dealing with her fear of attack.

She must accomplish all this quickly; it won't do to take all day. Then she must compare any single suitor with all other available males, in addition to considering her past choices and how they turned out.

"It turns out to be very similar to a search for job candidates," says Gail Patricelli. "First a résumé check, then a short interview, then a longer interview. Models from economics having to do with finding good job candidates (called the 'secretary problem' by the—obviously male— economists who made the models) turn out to predict female bowerbird behavior well." Each time a female bowerbird encounters a new male, she must compare him with her memory of previously encountered males and is more likely to accept the new one if he compares favorably.

BUT WHY IS SHE so choosy? Why bother to search for a male who's good at learning, decorating, mimicry, dancing, and problem solving?

One explanation is that females may use a male's bower the way female songbirds use male song, as a way to evaluate his overall genetic fitness, including his cognitive ability. The many display traits encode information the female needs to know about the male in order to judge his fitness as a mate—that he comes from a good egg, that he has no parasites, that he has stamina, fine motor skills, and superior cognitive abilities. According to Keagy and Borgia, it's in the totality of a male's display—bower, decor, song, and dance—that females may read his full worthiness as a sire, and perhaps especially his cognitive capacity. "All of these elements of the male display seem to have some component of cognition," says Borgia. "Each trait can tell a female something about the male," adds Keagy. "For instance, the number of blue decorations in-

dicates his competitive ability; the number of snail shells (which are dura-
ble and collected over years) offers information about his age and his
survival ability; his mimicry suggests his learning and memory capacity;
and his bower construction reveals his motor coordination and skill."
A single display trait is not necessarily a reliable measure. "So a female
uses all of these traits together to get a more accurate indicator of a male's
overall quality," Keagy explains. "It's like a sexually selected intelligence test of males complete with a total score, but also scores for different
categories." (As it happens, research suggests that human females do the
same, accurately assessing male intelligence by observing their behavior
on verbal and physical tasks: Intelligent human males, it turns out, are
more appealing.)

"To the extent that these things are important to females, then she's
choosing cognitively capable males," explains Borgia. But he cautions:
"One can debate to what extent cognition is something purposefully chosen by the female, or whether it's that males with better cognition can put
on better displays."

IN ANY CASE, a wise female satin bowerbird appears to seek males with
superior displays. Perhaps she picks so carefully so that her offspring will
inherit quality traits such as good health, a strong immune system, vigor,
and intelligence. It's called the good genes model. That's one idea.

There's another, more radical idea. Female bowerbirds and peacocks
and other choosy female birds may be drawn to gorgeous bowers and
displays because—well—because they're gorgeous. This was Charles
Darwin's *really* dangerous idea, says Richard Prum: that colorful feathers
or beautiful bowers can do two things at once. They can advertise desirable qualities like vigor and health. But they can also "just *be* desirable
qualities in and of themselves, without communicating any special meaning about fitness."

As Ronald Fisher suggested in his pioneering model of sexual selection, certain extravagantly beautiful traits—even when not useful—may

have evolved simply because they were preferred by the opposite sex. As Prum points out, Darwin's idea that female animals might appreciate beauty for its own sake was also daring for this reason. Males may gradually evolve beautiful traits, Darwin proposed—whether splendid feathers or song or bower—"through the preference of the females during many generations." Male peacock feathers, for instance, coevolved with the aesthetic sense of the female, who appreciates the glorious colors and patterns. In the case of the bowerbird, the beauty of the bower is shaped by the perception of the female. In other words, her mind shapes male display; she is the architect of the male bird's artistic creation and the brains required to achieve it, just as the female songbird is the architect of the male's elaborate song and the fancy neural networks that produce it.

IF THE FEMALE BOWERBIRD is indeed the artist who, by selection over long generations, helps to create these stunning bowers, it raises the question of how she may perceive beauty. Do bowerbirds have an aesthetic sense? Do they perceive beauty in the same way we do?

Shigeru Watanabe explores the thorny question of how another creature may experience aesthetics at his lab in Keio University in Japan. Some years ago, Watanabe tested the ability of birds to discriminate between human paintings of different styles—for example, cubist from impressionist. In the earliest such study, he trained eight pigeons to distinguish between the works of Picasso and Monet. The pigeons came from the Japanese Society for Racing Pigeons; the paintings, from photos of reproductions in an art book. The experimenters trained the pigeons to spot ten different Picassos and ten different Monets by rewarding them when they pecked at the pictures. Then they tested the birds with new paintings by the artists, never seen during training—and with paintings by different artists in the same style. Not only could the pigeons pick out a new Monet or Picasso, they could also tell other impressionists (Renoir, for instance) from other cubists (such as Braque). (This early work won the scientists an Ig Nobel Prize for "achievements that first make people laugh, then make them think.")

To probe whether birds might be capable of making distinctions based on human concepts of beauty, Watanabe trained pigeons to distinguish between "good" and "bad" paintings, as defined by human critics. He found that the birds could indeed pick out the beautiful from the ugly using cues of color, pattern, and texture.

All well and good, but do birds prefer any particular style of painting? To find out, Watanabe's team created a rectangular birdcage designed to resemble a hallway in an art gallery. Along the "hallway" were stationed screens showing different painting styles: traditional Japanese-style paintings by Ukioy-e, as well as impressionist and cubist paintings. The scientists timed how long the birds perched before each kind of painting. This time the art critics were seven Java sparrows. Five of the seven appeared to favor cubism over impressionism; six of the seven showed no clear favorite between the Japanese and impressionist paintings (perhaps a disappointment to the Japanese investigators). Still, this was the first study attempting to show that animals other than humans may have preferences in human paintings.

More recently, research has shown that distinguishing painting styles—using color, brushstroke, and other clues—is not at all unique to humans. Indeed, scientists have trained honeybees to tell a Picasso from a Monet.

This is easy stuff to poke fun at. The idea that birds and bees play favorites with human art looks a lot like anthropomorphism. But Watanabe's work is less about whether birds prefer Braque to Monet than it is about their powers of acute observation and discrimination of color, pattern, and detail.

Birds are visual creatures. They make quick decisions based on visual information from heights at great speed. Pigeons shown a series of landscape photographs taken successively can detect slight visual differences that are hard for humans to pick out. They can also recognize other pigeons by sight alone. So can chickens. Just because the powerful little central nervous systems of pigeons or bowerbirds are organized very differently from our own doesn't mean they're less capable of exceptional visual perception and fine discriminations.

Consider the business of evaluating subtle movements in dance: Some female birds, it turns out, are mind-bogglingly good at this—the golden-collared manakin (*Manacus vitellinus*), for instance, a bird known for its astonishing acrobatic courtship displays. In manakins as in bowerbirds, a male's chances of mating rely on female assessment of his display. Male manakins perform a "jump-snap" display. It starts with a leap between little saplings. Then, midjump, the bird flips its wings upward in a noisy wing snap. On touching down, it quickly swivels its body into a beard-up statuary position to show off its bright yellow beard, or throat feathers. It's a supremely difficult move, requiring exquisite neuromuscular coordination and great stamina. Think of a gymnast nailing a perfect landing.

Like bowerbirds, only a small number of male golden-collared manakins win the most matings. In the hopes of finding out what distinguishes the winners, a team of experimenters recently used high-speed cameras to record the displays of wild male manakins. They discovered that females prefer males that perform dance moves at a higher speed. But the difference between the beard-up *rond de jambe* of one male and the next is measured in mere milliseconds. "The capacity of females to discriminate slight differences in male-choreographed motor patterns (dances) has previously been shown only in humans," say the researchers.

I think I could distinguish a bad ballet dancer from a good one. But could I tell a 3.7-second *grand jeté* from a 3.8-second one? Somehow, the female golden-collared manakin registers these whiskers of temporal difference.

When scientists looked at the brains of both males and females of the species, they found specialized motor control circuitry in males and specialized visual processing circuitry in females. Further research into several different manakin species revealed a tight correlation between the complexity of a male manakin's display and the weight of his brain.

In birds, it seems, sexual selection for acrobatic motor behavior can drive the evolution of brain size. "The manakin brain," write the scientists, "has been shaped by evolution to support both male performance and female assessment of courtship display." More evidence for the mating brain hypothesis.

WHEN IT COMES TO artistry or display, birds are capable of making subtle visual distinctions, just as we are. But as scientists are quick to note, we must be careful to consider this in light of the bird's Umwelt, or sensory and cognitive world. Animals view the world with different sensory systems than our own. Color, for instance, is not a property of the physical world but the fabrication of whatever visual system may process and analyze it. Birds have possibly the most advanced visual system of any vertebrate, with a highly developed ability to distinguish colors over a wide range of wavelengths. We have three kinds of cone cells for color vision in our retinas; birds have four. Some species of birds are sensitive to the ultraviolet end of the spectrum, where we're blind. Moreover, in each of a bird's cone cells is a drop of colored oil that enhances its ability to detect differences between similar colors.

"We don't know if there are differences in how the bowerbird brain processes colors," says Gerald Borgia. "Our experiments with decoration color use in satin bowerbirds provide no evidence that they're seeing much differently than we are. Three species, however—the great, the spotted, and the western bowerbirds—may see into the UV part of the spectrum." In other words, a bowerbird's array of objects may look much the same to the bird as it does to us, or it may glow and sparkle in ways we can't imagine.

Still, some of the cues birds use to make visual judgments may be rooted in universal principles of beauty—or at least attractiveness—such as symmetry, pattern, and contrasting colors. Experiments in the 1950s, for instance, showed that crows and jackdaws have a distinct preference for regular, symmetrical patterns.

NOBEL LAUREATE KARL VON FRISCH once wrote, "Those who consider life on earth to be the result of a long evolutionary process will always search for the beginning of thought processes and aesthetic feelings in animals, and I believe that signature traces can be found in the bowerbirds." Given the shared biology in the nervous systems of birds and humans, wouldn't it be wrong to presume that there is nothing in common between our aesthetic sense and theirs?

When I asked Gerald Borgia if he thought bowerbirds might have an aesthetic sense, a special feel for the beautiful, he said he had no idea. "Over time, the birds seem to be able to build up a kind of image of what their decoration display should look like," he told me. "It's usually older birds that seem to have this, and then when the younger birds take over the bower, they don't appreciate what's there." Case in point: After that spotted bowerbird who collected stained glass died, another spotted bowerbird appropriated his bower. "But the newcomer just piled up the shards," says Borgia. "He didn't seem to know what to do with them."

As to whether this indicates an aesthetic sense in the older males: "I find that terminology slippery, so I try to avoid it," says Borgia. "I know what beauty looks like to me. I think their bowers are beautiful, but I don't know if that's the reason they built them that way."

It's true. We haven't the slenderest notion of what a male bowerbird really thinks of his display. But we do know that he does not waste time or make a fool of himself chasing after the ladies. Instead, he gathers up the blue objects of his world and lays them out. He designs. He builds. He sings. He dances. The female, with great acuity, marks his efforts. Sharp, attentive, creative? If she likes what she sees, she offers him her body. And so it goes.

seven

A MAPPING MIND

Spatial (and Temporal) Ingenuity

Imagine you're somewhere in Canada driving south toward the lower forty-eight states. It's late fall, and you're heading to a cottage in warmer climes hundreds of miles down the road. Suddenly, you're plucked from your car, bundled into a sealed vehicle, and taken to an airport. The next thing you know, you're jetting across the country blindfolded, with no idea where you're going. After several hours in the air, you land, and you're immediately hustled into another sealed van and taken to an unknown location. When you're finally freed, none of your surroundings is even vaguely familiar. You have no access to GPS, no map or landmark or compass. But you must nevertheless find your way back to that cottage you were heading for, though it's now somewhere across the country, thousands of miles away.

How would you fare?

That's pretty much what happened to a flock of white-crowned sparrows not long ago. These little songbirds with their crisp black-and-white-striped crowns, each a single feathered ounce of fortitude, normally migrate from their breeding grounds in Alaska and Canada to their wintering grounds in Southern California and Mexico. One day when a flock of sparrows was passing through Seattle on their way south, researchers captured thirty of the birds, fifteen adults and fifteen juveniles. They packed the sparrows into crates and flew them by small aircraft all the way across the country, twenty-three hundred miles from their normal migratory flight route, to a release site in Princeton, New Jersey. There they let the birds go to see if they could find their route back to their wintering grounds. Within the first few hours of release, the adult sparrows had reoriented themselves and set off traveling solo across the country, aiming directly at Southern California and Mexico. Even the youngest adults, who had made only one migratory journey in their brief lives, found their bearings and headed for their winter home.

A WHITE-CROWNED SPARROW may be nut brained, but it's far more gifted at navigating than most modern humans. We have our mental maps, all right, created by learning the relationships between familiar landmarks—say, where the post office or the bakery sits on a familiar street grid. But this is something else. That a sparrow transported far beyond its known territory seems to know exactly how to get back on course is one of the astonishments of the bird mind.

A good memory doesn't explain it. And neither do theories that focus solely on instinct or eyesight or magnetic cues or sensitivity to the azimuth of the sun. As Julia Frankenstein, a psychologist at the University of Freiburg's Center for Cognitive Science, writes, "Navigating, keeping track of one's position and building up a mental map by experience is a very challenging process." It involves such cognitive skills as perception, attention, calculating distances, approximating spatial relations, decision making—all very hard work, even for our big mammalian brains.

How do birds do it?

The ability was once thought to be innate in birds, a matter of instinct. Now we know that bird navigation involves sensing, learning, and, above all, a remarkable ability to build a map in the mind, one far bigger than we ever imagined and made of strange and still mysterious cartography.

MUCH OF WHAT WE know about how birds find their way we've learned from a humble species, a bird that for hundreds of years has been subjected to a version of the experiment endured by those white-crowned sparrows: the pigeon race. Sometimes known as the "poor man's horse race," pigeon racing involves first training the birds by releasing them from a basket at locations they've never seen at greater and greater distances from their lofts. Eventually the pigeons can be relied on to return from distances up to a thousand miles, beelining homeward over wide stretches of unknown country at speeds averaging fifty miles per hour. Most make it home—though not all.

Take the story of Whitetail.

One April morning in 2002, pigeon racer Tom Roden saw a bird with a pale tail flutter to a landing in his loft at Hattersley, Hyde, near Manchester, England. The pigeon looked vaguely familiar. Roden, a long-time pigeon fancier and racer, checked the registration ring on the bird's leg and realized it was his own racing bird, which had vanished five years earlier during a race across the English Channel.

Whitetail's disappearance had been something of a riddle. He was no ordinary bird—in fact, he was a champion, winner of thirteen races and veteran of fifteen channel crossings. But this had been no ordinary race. Its catastrophic failure earned it the moniker the Great Pigeon Race Disaster.

The race was held in honor of the Royal Pigeon Racing Association's centenary celebration. Early on a Sunday morning in late June 1997, more than sixty thousand homing pigeons were released at a field near Nantes in southern France to fly home to their lofts all over southern England. At 6:30 A.M., the air whistled with the sound of pigeon wings as the birds whooshed into the sky for the journey north four hundred to five hundred miles. By eleven A.M., the vast majority of the feathered racers had flown two hundred miles to the edge of France and launched out over the English Channel.

Then something happened.

In the early afternoon, fanciers in England waited at their lofts for the fastest birds to arrive. But as the hours ticked by with no sign of wings in the sky, the disappointed sportsmen scratched their heads in bafflement and dismay. Finally, a few birds straggled in, including some of Roden's racers, the slowest in his loft. But no Whitetail. The champion bird, along with tens of thousands of other experienced racers, never made it home that day. The reason for their disappearance remains mysterious, though clues have emerged (which I'll return to later).

Fast-forward five years to that cool April morning. Roden had just stepped out to walk his dog when he spotted Whitetail. "I was absolutely amazed," he told the *Manchester Evening News*. "I always said I thought Whitetail would one day make his way back . . . but even I'd given up hope of seeing him again."

THE GREAT PIGEON RACE DISASTER was notable because of its rarity. Racing pigeons seldom go astray; the vast majority make it back to their lofts—even when the distances are great. A good example is Red Whizzer Pensacola, a beautiful red checker cock, opal with a ruby bib and eye, who arrived at his home in Philadelphia after traveling 930 miles from a release site in Pensacola, Florida. The *New York Times* reported that the distance was the greatest yet covered by a homing pigeon either in this country or abroad. The winning bird was bestowed with a golden leg band inscribed with his loft and register number and retired from his duty as a flyer.

That was in 1885. Homing pigeons have since repeated this feat—and much greater ones—thousands of times, in races around the world. Calamitous losses, called smashes or busts, do occasionally occur. A year after the English Channel disaster, for instance, thirty-six hundred homing pigeons were released in races in Pennsylvania and New York and only a few hundred made it home. No one knows why.

Is it surprising that racing pigeons occasionally "come a cropper," as homing pigeon expert Charles Walcott puts it? Or is the bigger wonder how they normally find their way home from places they've never been? It may seem obvious that a bird would remember yesterday's route to that field full of grubs or the way back to the warm, dry hollow of its nest hole. But finding a route home over hundreds of miles is another matter.

Even the remarkable journeys of homing pigeons pale in comparison with the jaw-dropping voyages of long-distance migratory birds—lately illuminated by new technology. Birds wearing tiny geolocator backpacks have revealed the details of their marathon migrations. The tiny blackpoll warbler, a bird of boreal forest, leaves New England and eastern Canada each fall and migrates to South America, flying nonstop over the Atlantic to its staging grounds in Puerto Rico, Cuba, and the Greater Antilles—a flight of up to seventeen hundred miles—in just two or three days. The Arctic tern, a bird who lives by his love of long daylight and bent for high mileage, circles the world in orbit with the seasons, flying from its nesting grounds in Greenland and Iceland to its wintering grounds off the coast of Antarctica—a round-trip of almost forty-four thousand miles. In an average thirty-year lifetime, then, a tern may fly the equivalent of three trips to the moon and back.

How in the world does it find its way? How does a red knot resting at Cape May on its spring journey north from Tierra del Fuego know how to pinpoint last year's breeding grounds in the distant northern Arctic? How does a European bee-eater traveling south from its summer season in a farm field in Spain find a course over the Sahara to its familiar patch of West African forest? How does a bristle-thighed curlew or a sooty shearwater steer homeward over a vast and featureless expanse of sea?

As one who gets easily lost in a small patch of woodland, I'm in awe of the navigating abilities of birds. How can they accomplish a feat few humans can carry out even with the help of a compass?

THE DOMESTIC PIGEON, *Columba livia*, is a good bird for probing such questions. Pigeons have a bad rap as gutter birds, rats with wings, pecking savagely at bread crumbs beneath park benches or rummaging in our city dumps. They're considered by some to be as dumb as a dodo (a close relative, in fact).

It's true that the forebrain of a pigeon has only half the neural density of a crow's forebrain. It's also true that pigeons may fail to realize that an egg or a squab is theirs unless it is immediately beneath them. They will accidentally trample their young to death or toss them from the nest. (Although, as one pigeon expert points out, "The squabs are so small and the pigeon's feet so comparatively large that the wonder is that more squabs are not tramped to death.") Pigeons are also notoriously inefficient nest builders, carrying only a single twig or coffee stirrer at a time, whereas sparrows will ferry two or three. And if a bit of nesting material drops midair, a sparrow will swoop down to catch it, while a pigeon will let it fall and fail to retrieve it.

So by some measures, yes, they may seem dim-witted. But in truth, they're far more bookish than you might imagine. They're handy with numbers, for instance, capable not only of counting (which, granted, a lot of animals can do, including bees) but also of grasping the arithmetic of loss and gain and learning abstract rules about numbers, abilities on par with primates. They can, for example, put images picturing up to nine objects in proper order from lowest to highest number. They can also determine relative probability.

In fact, pigeons are better than most people—and even better than some mathematicians—at solving certain statistical problems: the Monty Hall Dilemma, for instance, named for the host of the old television game show *Let's Make a Deal*. In the original game show, a contestant tried to guess which of three doors (displayed by the "lovely Carol Merrill") concealed a grand prize, such as a car. The other two doors harbored a booby prize, such as a goat. After the player chose a door, one of the remaining doors was opened, revealing no prize. The contestant was then given the option of staying with the initial choice or switching to the other unopened door.

In a laboratory version of the game, pigeons solve the puzzle successfully—picking the right "door"—more often than humans. Most human players opt to stay with their first choice despite the fact that switching doors would double their chances of winning. Pigeons, in contrast, learn from experience and go with the odds, switching their choice.

The puzzle appears to defy logic. It would seem that with two unopened doors remaining, your chances would be fifty-fifty that the prize was behind one of them. But in truth, switching doors will give you a 66 percent chance of winning. Here's why: The probability that you picked the right door initially is one in three. So there's a two-in-three chance that you picked the wrong door. When Monty opened the door with the goat, those odds remained. (Monty always knew where the car was and wouldn't open that door.) This means the other door had a two-thirds chance of being the right door. I know. I still have trouble getting my head around this. So do many mathematicians. (When the Monty Hall Dilemma appeared in the "Ask Marilyn" column of *Parade* magazine, along with the correct solution, Marilyn vos Savant received more than nine thousand letters disagreeing with her solution, many of them from mathematicians at universities.) But apparently, pigeons do not. Initially the birds choose randomly, but eventually they learn to switch. Their successful approach to the problem calls for the use of empirical probability—that is, observing the outcomes of numerous trials and adjusting one's behavior accordingly to win a reward. More often than not, pigeons adopt optimal strategies for this, maximizing their chances of winning, whereas humans fail to do so, even after extensive training.

Pigeons are also good at making distinctions as to whether sets of objects are identical or not—a skill the American psychologist William James once called "the very keel and backbone of our thinking." They certainly aren't the champions of this skill. That honor may belong to Alex, the African grey parrot that Irene Pepperberg so brilliantly studied before his demise in 2007. Alex was not only nearly flawless in stating whether two objects were the same or different with respect to color, shape, or material, but could say "none" if there were no similarities or differences. He could also categorize more than one hundred objects based on these characteristics.

Still, pigeons do very well at distinguishing arbitrary visual stimuli such as letters of the alphabet and—as we know—the paintings of Van Gogh, Monet, Picasso, and Chagall. They can differentiate between photographs that contain human beings (whether they're clothed or naked) and those that do not. They're highly skilled at recognizing the identity of a human face and even reading its emotional expression. They can learn and recall more than one thousand images, storing them in long-term memory for at least a year.

And to the point here: They're a far cry better than we are at finding their way in the world—without the benefit of technology. Because of this, they have found themselves in the position of "lab rat with wings," at the pointy end of investigations trying to fathom the mystery of how bird navigation works.

LATELY I'VE BEEN PONDERING the pigeons that flock together like hooded monks or tourists on the bricks of our downtown public spaces. The more I see of these birds, the more I like them. They may be shy, nervous of novelty. But they're also scrappy and adaptable. Up close, their crops are iridescent with rainbow.

Thanks to breeding since ancient times, there are dozens of pigeon varieties. Fancy ones like tumblers, priests, nuns, fantails, and Dragoons have been bred for appearance and show, sometimes to extravagant effect. (The pouter, for instance, has been aptly described as looking like a tennis ball stuffed into a glove.)

Homing pigeons are bred to home and to race. The typical feral ones you see in American cities descend from escaped tame immigrant homing pigeons brought to the nation on ships with European settlers in the early 1600s—the first exotic birds to arrive on these shores.

The city slickers I see do a lot of walking, swaying their stubby bodies back and forth like ducks, and sometimes pulling themselves upright to strut with crisp little soldierlike marching steps. They seem wary of perching in trees, instead beading our telephone wires with their plump

bodies, or tucking into the architectural nooks and crannies, the capi-
tals, abutments, girders, corbels, and scrolls of city bridges and buildings,
tails mashed vertically against the wall. This penchant for skimpy ledges
has always struck me as an odd predilection—and uncomfortable.

Why would the common pigeon shy away from high arboreal perches
and favor narrow ledges? Because it descends, like all domestic pigeons,
from the wild rock dove, a bird that nests in the sea cliffs and rocky islands
of the Mediterranean. Rock doves forage for seeds in nearby fields and
then return home with food for their squabs. It was in this context that the
doves' natural ability to navigate homeward probably evolved.

HUMANS HAVE EXPLOITED the pigeon's homing instinct for at least
the past eight thousand years. That's according to the bible of dove lit, *The
Pigeon*, first published in 1941 by Wendell Mitchell Levi, a pigeon fancier, scientist, and first lieutenant in charge of the pigeon section of the
United States Army Signal Corps during World War I.

"Wherever civilization has flourished, there the pigeon has thrived,"
Levi writes, "and the higher the civilization, usually the higher the regard
for the pigeon."

Through the centuries, homing pigeons have been used as fleet messengers, couriers, and spies—by ancient Romans to announce victories
in the Coliseum, by Phoenician and Egyptian sailors to herald the arrival
of ships, by fishermen proclaiming catches, and by bootleggers passing
word between ship and land bases during Prohibition. It's said that the
Rothschild bank learned early of Napoleon's defeat at Waterloo by pigeon
courier and shifted its investments. In the mid-nineteenth century, Paul
Julius Reuter launched his news service with pigeon posts carrying stock
prices between Aachen and Brussels. And in the early twentieth century,
pigeons carried messages of safe arrival or distress from boats traveling
between Havana and Key West, Florida.

During both world wars, pigeons were used for the quick conveyance
of intelligence. The birds were suited up with ciphered papers and sent

across enemy lines to relay news of troop movements or communicate with resistance workers in occupied countries. These winged spies bore names such as The Mocker, Spike, Steady, The Colonel's Lady, and Cher Ami, who, according to Levi, completed her mission "despite a broken leg and breastbone suffered en route." There was a bird named President Wilson, who lost his left leg in the Great War. And Winkie of Scotland, who crashed with his bomber crew in the North Sea. Winkie was released from the ruined aircraft, and in a twinkling flew 120 miles to his loft near Dundee, alerting the air base there, which sent rescue aircraft to save the marooned bomber crew.

At its peak in World War II, the U.S. Pigeon Service possessed fifty-four thousand birds. "We breed for intelligence and stamina," explained one handler. "What we want is a bird that will get back, one that won't get flustered, one that is intelligent enough to be self-reliant. Now and then we get dumbbells, of course. You can spot them early. They don't know enough to get back in the loft or they sit in a corner and sulk." But most pigeons, he said, are "intelligent. Highly intelligent."

Among the most celebrated of these winged messengers was G.I. Joe. Dispatched by the British to abort a scheduled bombing of a German-held town because a brigade of a thousand or more British troops was already occupying it, Joe made the 20-mile flight in twenty minutes, halting the bombers just as they were warming up for takeoff. Then there was Julius Caesar, a blue checker splashed cock, who was parachuted out of Rome and released in southern Italy, where he took off in a southerly route to his loft in Tunisia with vital information for the North Africa campaign. And Jungle Joe, a gallant four-month-old bronze cock, who flew 225 miles against strong wind currents and over some of the highest mountains in Asia to deliver a message that led to the capture of large parts of Burma by Allied troops.

Officials in Cuba still use the birds to transmit election results from remote mountainous areas, and the Chinese have lately built a force of ten thousand messenger pigeons to deliver military communications between troops stationed along their borders, in case of "electromagnetic interference or a collapse in our signals," as the officer in charge of the pigeon army explains.

"IT IS FREQUENTLY AFFIRMED that the carrier pigeon finds its way without the exercise of intelligence or observation, and merely by the aid of some incomprehensible instinct," wrote Charles Dickens in 1850. "But from my own observations . . . I am convinced this is a mistake."

Darwin, Dickens's contemporary, proposed that pigeons might somehow register the snaking route of their outward journey and then use this information to figure out their homeward route. Now we know that's not so: Even pigeons who have traveled a circuitous route inside rotating drums in a sealed vehicle can make their way home from an unfamiliar point of release—not by retracing a route but by more or less direct flights.

Returning to a known location over familiar ground is one thing. True navigation is something else. It's the ability to choose the right direction to a goal from an unfamiliar place using only cues present locally, not those gleaned by having flown the route before. We rely on technology for this: GPS and map software that tell us exactly where we are anywhere on earth, and how to get from where we are to where we want to go. Birds, it seems, have their own internal positioning systems, which, like GPS, may indeed be global.

To see if birds are truly navigating, scientists put them on boats and planes and drive them around in cars (like those hostage sparrows) to land them in a distant, unfamiliar place with no clue of distance or direction. Then they release the birds and watch how they reorient. It's called a displacement study, and it's a powerful tool to investigate true navigation.

Scientists suspect that pigeons and other birds navigate using a two-step "map-and-compass" strategy. First, they determine where they are at the point of release and which way they need to travel to get home. (This is the map step: In human terms, it's the spatial coordinate system that suggests "I'm south of home, so I need to travel north.") Then they use landmarks or celestial or environmental directional cues as a compass to keep them on the straight and narrow. The whole system, including both map and compass, appears to consist of multiple elements involving different types of information—sun, stars, magnetic fields, landscape features, wind, and weather.

The compass part is fairly well understood, due in large part to thousands of studies depriving birds (often pigeons) of one sense or another, displacing them, and then seeing whether they're thrown off course.

Pigeons, like humans, are eye-minded creatures. It would be surprising if they didn't use that grove of gnarled oak trees, that oxbow curve in the river, that hedgerow or weird triangular skyscraper, to home in on their lofts. And, it turns out, they do—at least on the very last leg of their journey.

The sun helps, too. Like bees, pigeons use the sun as a compass with the help of a precise little internal clock possessed by all birds. The internal clock gives them a sense of time, so that at any point in the day they know where the sun should be. But to use the sun as a compass in navigation, a young pigeon must learn its path. She does this by observing the sun's arc at different times of day, learning how fast it moves—about 15 degrees per hour—and internalizing a representation of the arc. If she's exposed to the sun only in the morning, she can't use it to navigate in the afternoon. She calibrates her sun compass daily, perhaps using the polarized light visible near the horizon at sunset. Once she masters this use of sunlight, she favors it over other cues. Even within a couple of miles of her home loft, she will rely not on familiar landmarks but on her sun compass.

But here's the wonder: Even pigeons whose vision is masked with frosted lenses can orient homeward, all the way to their loft. According to Charles Walcott, a professor emeritus of ornithology at Cornell University, when birds with frosted lenses approach their loft, they come in high and sort of "helicopter" down. Something else guides them.

MORE THAN FORTY YEARS AGO, William Keeton of Cornell University showed that under overcast conditions, pigeons fitted with little magnetic bars become disoriented and home more slowly than controls. (Lest you think that's because we might all flounder with a barbell bound to our backs: The controls wore nonmagnetic brass "dummy" bars.)

Earth is like a giant magnet: Magnetic lines of force, or field lines, emanate from its poles, weakening and flattening as they near the equator. Birds seem to be able to detect even tiny changes in the inclination, or vertical angle, of the magnetic field and may use these to determine their latitude.

The first hint that magnetic fields might guide birds in their journeys came from experiments in the late 1960s with caged European robins. The birds were kept in rooms isolated from any outdoor environmental cues. European robins normally migrate south from northern Europe to southern Europe and Africa. During their period of migratory restlessness, called *Zugunruhe*, the captive birds—their hearts racing as if to power flight—consistently seemed to want to escape toward the south, though they had no visual clues as to where south might be. When scientists wrapped their cages in electromagnetic coils, the birds were confused and shifted the direction of their fluttering and hopping.

Lots of creatures, from bees to whales, perceive magnetic fields and use them to orient. However, we're still not certain how animals sense the fields. Detecting them with sensitive electronic instruments is one thing. But "sensing magnetic fields as weak as that of the Earth is not easy using only biological materials," says Henrik Mouritsen, a biologist who studies the mechanisms underlying animal navigation at the University of Oldenburg in Germany. Birds possess no obvious sense organ devoted to the task. But because the field can pervade tissue, the sensors may be hidden deep within their bodies.

One model holds that birds "see" magnetic fields with special molecules in the retina activated by certain wavelengths of light. Magnetic signals seem to affect the chemical reactions of these molecules, either speeding them up or slowing them down, depending on the direction of the magnetic field. In response, the retinal nerves fire signals to the visual areas of a bird's brain, making it aware of the field's direction. It all occurs at the subatomic level, involving the spinning of electrons, which suggests something extraordinary: Birds may be capable of sensing quantum effects. The sensing seems to involve a part of the forebrain linked to the eyes, known as cluster N. If cluster N is damaged, birds can no longer sense which way is north.

What would they actually see? It's hard to know. Perhaps a ghostly pattern of spots, or of light and shade, that would remain in place as the bird moves its head from side to side.

A second theory suggests that a magnetic sensor made of tiny crystals of iron oxide—kind of like a compass needle—may be lodged somewhere in a bird's body. This sensor would detect gradients in a magnetic field and translate them into a neuronal impulse.

Not long ago, scientists thought they had found just such minute magnetic sensors in the beaks of pigeons, specifically, in six clusters of iron-rich cells they found in the nasal cavity of the birds' upper beaks. But when researchers looked further, poring over 250,000 tissue slices from the beaks of almost two hundred pigeons, something didn't add up. The number of these iron-bearing cells varied widely from bird to bird. One pigeon had only 200; another, more than 100,000; and another, with a beak infection, had tens of thousands located smack-dab at the site of the infection. The iron-rich cells appear not to be sensory cells at all but white blood cells known as macrophages, which are simply recycling iron from the red blood cells they engulf.

End of story? Not exactly. New evidence suggests that magnetoreceptors somewhere in a bird's upper beak close to the skin are involved in recording magnetic intensity, which varies with latitude. Severing the nerve that links a bird's beak with its brain sabotages its ability to establish its position. But just what is detecting the magnetism—and where in the beak it exists—remains a mystery.

To confound the picture, another possible niche for magnetoreceptors emerged recently: this time, in tiny balls of iron found inside hair cells— sensory neurons within a bird's inner ear—suggesting that birds may "hear" magnetic fields. However, removing the inner ear of homing pigeons has no effect on their ability to home.

Wherever the sensor may be, it appears to be extraordinarily sensitive. In 2014, Mouritsen and his team reported in *Nature* that even extremely weak electromagnetic "noise" generated by human electronic devices in urban environments may disrupt the magnetic compasses of migrating European robins. We're not talking cell towers or high-voltage transmission lines here; more like the background buzz of everything run by electrical currents. This news caused some shock waves in the scientific world. If it's true, this kind of "electro-smog," as it's known, may already be causing birds navigational problems serious enough to affect their survival.

For a long time scientists thought that a bird's magnetic compass was just a kind of backup system for overcast days. Far from it. Together with the sun compass, it's essential to their navigational system. So maybe birds harbor different types of magnetoreceptors that work together, allowing them to sense even the tiniest fluctuations in magnetic fields. Parsimonious as birds may be, perhaps in this department they go a little overboard—so that, for instance, a pigeon flying over the Mediterranean Sea on a moonless night might find its way to a dovecote in North Africa.

SO MUCH FOR the compass piece of the puzzle. To navigate, a bird also needs something akin to a map to determine its position at the start of its journey—and where that space is in relation to where it's going so it can head in the right direction. Do birds have such a thing? A map inside the mind?

The idea goes back to the 1940s, when Edward Tolman, a psychologist at the University of California, Berkeley, first proposed that mammals might possess a "cognitive map" of their spatial environment. Tolman observed that rats in special mazes were able to figure out new, more direct routes or shortcuts to destinations that held a food reward. "In the course of learning," said Tolman, "something like a field map of the environment gets established in the rat's brain," indicating routes, paths, blind alleys, and environmental relationships, which the rats may later call on. (Those who pursued Tolman's vein of cognitive map research were affectionately known as Tolmaniacs.)

Tolman proposed that humans, too, build such cognitive maps, and bravely suggested that these maps help us navigate not only space but the social and emotional relationships in "that great God-given maze that is our human world." A narrow-minded map can lead one to devalue others and in the end to "desperately dangerous hates of outsiders" ranging in expression "from discrimination against minorities to world conflagrations," Tolman wrote. The solution? Create broader cognitive maps in the mind that encompass bigger geographical boundaries and a wider social scope, embracing those we might consider "other," and in this way encourage empathy and understanding.

THE DISCOVERY THAT BIRDS might make mental maps of their physical surroundings—if not their social and emotional ones—came from putting pigeons to the same sort of maze tests Tolman used. Like rats, it turns out, pigeons have an excellent memory for spatial information; they remember landmarks they've visited before—how far apart they are and in what direction they lie—and they use this information to guide them to new locations.

It's called small-scale navigation, and some birds are very good at it, indeed. The champs are those "scatter-hoarding" birds such as Clark's nutcrackers and western scrub jays. These members of the crow family are masters of the spatial memory game on a colossal scale.

Clark's nutcrackers (*Nucifraga columbiana*), light gray, crow-shaped birds with handsome black wings, are nicknamed "camp robbers" for their habit of scrounging in campsites. They are native to the Rocky

Mountains and other high regions of western North America. To survive the harsh winters there, a single nutcracker will gather more than thirty thousand pine seeds in a single summer, carrying up to one hundred seeds at a time in a special large pouch under its tongue. These it buries in up to five thousand different caches scattered throughout a territory of dozens, even hundreds, of square miles. Then later it finds the scattered treasures. Nutcrackers recall the locations of their own individual stashes and will go directly to them without expending a lot of energy looking elsewhere. They rely almost completely on memory to locate their personal caches—and they can remember them for as long as nine months, despite radical changes in the appearance of the landscape across the seasons caused by snow, leaves, or shifting rock and soil.

A pine seed is tiny, and so is each of the caches. The bird digs for his treasure with a very small shovel indeed, his daggerlike bill, and striking his target demands precision measured in millimeters. Even the slightest error in recalling the location of a cache might mean that it's never found. Seven times out of ten, a Clark's nutcracker nails it. (A particularly humbling statistic when I consider my own inability to keep track of, say, my car keys or where I've planted my tomato seeds.)

The question is, how do they find the seeds once they've cached them? Olfactory cues play no part. One theory holds that they create a mental map of big, tall landmarks, such as trees and rocks, that won't get buried by snow. They register and remember the location of caches relative to these landmarks, using distance, direction, and even geometric rules and configurations. For instance, they might register that a cache site is situated halfway between two tall landmarks—or at the third point of a triangle created between the two landmarks and a target location. Imagine recalling five thousand such locations.

WESTERN SCRUB JAYS—those masters of social trickery—remember not only *where* they stashed their caches (and who was watching) but also *what* they stashed there and *when*. This is important because the scrub jay squirrels away not only nuts and seeds but fruit, insects, and worms, foods that perish at different rates. Cached insects can spoil in days if the temperatures are high enough, while nuts and seeds can last for months. A series of creative experiments by Nicola Clayton and her team at Cambridge University showed that the birds retrieve the more perishable food before it rots, leaving the nonperishables, such as nuts and seeds, until later. The jays use their experience of how quickly food degrades to guide their choice in recovering their caches. Remembering that perishable food items may need to be retrieved sooner requires recalling cache locations, cache contents, and the time of caching. This ability to remember the what, where, and when of specific past events is thought to be akin to human episodic memory, the remarkable capacity to remember specific personal experiences. Like us, the birds seem to be using events that happened in the past (what they buried when) to figure out what to do now or in the future (dig up or save until later).

Clayton and her team have followed up on these experiments with others strongly suggesting that the scrub jays also seem capable of some degree of planning, or at least forethought, providing them with flexibility to act in the present in such a way as to boost their future chances of survival.

To probe whether scrub jays may plan for the future, Clayton and her colleagues housed eight jays in large cages with access to two different compartments. The first compartment always held breakfast, and the second compartment did not. The birds went without food overnight, and then in the morning were moved to one of the two compartments. After three mornings of exposure to each compartment, the birds were given food in the evening, pine nuts, which they were allowed to eat their fill of, and then store the surplus in either compartment. The jays cached their nuts in the "no breakfast" room—presumably in anticipation of their hunger there the next morning.

The researchers then added a twist. They offered the jays different foods in each compartment—peanuts in one, and dog kibble in the other.

This time when the birds cached their surplus, they distributed the food so that each room had equivalent offerings.

In subsequent experiments on Eurasian jays, Clayton and her colleague Lucy Cheke showed that the birds store the specific food they will want in the future (the one they haven't been fed recently), apparently planning for their impending needs while ignoring their current desires. "Whether jays 'pre-experience' the future remains an open question," write the researchers, "but our results provide strong evidence that they can act for a future motivational state that is different from their current one, and do so flexibly."

This work suggests that some birds, at least, seem capable of the two essential components of mental time travel—the ability to peer back into the past (what was I fed where?) and forward into the future (what will I be hungry for tomorrow and where should I stash my food?)—once considered unique to humans.

BUT BACK TO the spatial genius of western scrub jays. There's more. As we know, scrub jays steal one another's caches. Remarkably, cachers can recover caches that have been moved and those that haven't with equal precision. A thieving jay, for its part, exploits some sophisticated mental mapping of his own. It relies on spatial memory to locate the food it witnessed being cached by another bird, and it can recall the particular spot even if it's watching from a distance and has to mentally rotate the location in its head.

HUMMINGBIRDS APPEAR TO POSSESS a similar small-scale navigational genius.

Each spring my friend David White hangs a nectar feeder on a bungee cord with an S-shaped hook in his yard in central Virginia. Between seasons he takes down the feeder so the raccoons don't get it, but leaves

the cord and hook so he can easily hang it up again come April. Sometimes he forgets to reinstall the feeder. But to his delight, the ruby-throated
hummingbirds remind him, showing up around April 13, a day or two
before he usually rehangs the feeder, and hovering around the empty
S hook. The hummers know where to be—and when.

I've watched these nectarivores whiz through my window boxes in
spring, shuttling back and forth between flowers like whirring tops—
energy made visible, their wings a gauzy blur. A ruby-throated hummingbird weighs about 3 grams, less than an old penny.

The rubies that thrum and pivot around my plants seem not to buzz
the same blossom twice. Does this mean they carry a map in their heads
of the flowers they recently emptied and those that still carry nectar? (Or,
in the case of David's hummers, the location of all the hanging feeders in
a neighborhood?)

Keeping in mind the handful of blossoms in my window box is one
thing. Remembering the thousands of flowers that make up a typical territory for a hummingbird is another. But it makes sense that these birds
would devote brainpower to this sort of energy-saving strategy. Hummingbirds lead very energetically expensive lives. Not only does their
rapid wing beat of up to seventy-five times a second suck up calories;
so do their high-speed chases of rivals and their diving, waggling, zigzagging shuttle flights to attract mates. To fuel their air derbies, they
have to harvest hundreds of flowers per day; they don't want to waste a
dime visiting blossoms they've already sucked dry. So they keep track.
And they do it, apparently, not on the basis of color or shape or other visual tips offered by the flowers themselves, but rather through spatial
cues, as food-storing jays and nutcrackers do.

Sue Healy of the University of St Andrews probes the cognitive abilities of hummingbirds in the wild. She studies the rufous hummingbird,
a tiny bright orange bird known for its feisty, pugnacious defense of the
flowers on which it feeds. Healy's recent work suggests that these pen-
nyweight wonders can register the spatial location of a flower or feeder
in a large featureless field by visiting it once for only a few seconds. And

they can return to that location with impressive accuracy, even if the flower itself is absent. Moreover, they keep track of the nectar quality and content of individual flowers and their refill rates, revisiting blooms only after the flowers have had time to restock.

It's still a mystery what spatial cues they use to zero in on their quarry. Healy's research suggests that they rely on landmarks as a kind of scaffold for their mental map, as food-caching birds seem to do. But it's not a simple matter. In Healy's observations, nearby landmarks "were (to our eyes, at least) remarkably uniform: the ground was quite flat, and covered by vegetation." The more distant landmarks, however—trees rimming the field, three-thousand-foot mountains framing the valley—were very visible from all points in the field. But it's not clear how the birds could use such large landmarks to so accurately pinpoint where particular flowers or feeders are—or where they should be.

SCIENTISTS HAVE ASSUMED that homing pigeons possess maps like these in their heads, dotted with different memorized locations, just on a bigger geographical scale. But no one had really tested the idea outside the laboratory until recently, when Nicole Blaser (then a PhD student at the University of Zurich) devised an inspired experiment.

Blaser wanted to show that pigeons navigate not by means of a simple, robotic response to environmental cues, but rather with the help of a genuine navigational map in their brains, which allows them to choose different goals and the best routes to reach them.

If a pigeon is a sort of "flying robot," navigation should be a relatively simple two-step process: Compare an environmental cue such as a magnetic signal at an unfamiliar location with the same cue at a familiar one, like the home loft. Then move in a direction that systematically reduces the gradational difference between the two cues. This robotic "loftocentric" strategy, as Blaser calls it, would mean that the birds memorize only one location (the home loft), then navigate back to it following various gradient differences in environmental cues until they reach home.

How to show that pigeons carry in their heads a genuine map of multiple locations?

Blaser decided to give a flock of 131 pigeons a choice of where to fly: to a home loft or to a food loft, based on how hungry they were. First she trained all of her pigeons to recognize the location of a food loft. Every day, she ferried them by car to the food loft for regular feedings. (Pigeon research can be very labor intensive.) Then she released them from their home loft at gradually greater distances from the food loft, and vice versa, until they could fly efficiently from one loft to the other.

After the training, she took them to a completely unfamiliar place equidistant from both lofts, within twenty miles. She fed half the pigeons; the other half she left hungry. Then she released all the pigeons. Those with full bellies flew back to the home loft, but the hungry ones made for the food loft. They detoured only to get around topographical obstacles, two lakes and a mountain range, then corrected their courses. Not one hungry pigeon traveled by way of the home loft.

If the birds were navigating with a robotic "loftocentric" strategy, says Blaser, they would have oriented homeward first until they reached familiar terrain, then changed their flight path toward the food loft.

Flying directly to the spot that will satisfy their hunger is revealing on two counts, says Blaser: First, it shows that the birds are capable of making choices between targets according to motivation—a cognitive ability in and of itself—and also, that they're holding in their heads a genuine cognitive map, which includes knowledge of their own unfamiliar position in space relative to at least two known places.

WHERE IN THE DIMINUTIVE landscape of a pigeon brain could such a map reside?

In the same place it lives in our brain: the hippocampus, that neuronal network that helps us orient in space. We know this thanks in part to the effort of one Tolmaniac, anatomist John O'Keefe, who won the 2014 Nobel Prize for a remarkable discovery while doing his own maze studies with rats in the 1970s. In studying brain activity during the rats' maze running, O'Keefe and psychologist Lynn Nadel observed that certain special cells in the hippocampus fired only when the rats were in a specific place. As a rat meandered through a maze, these "place cells" fired in a spatial pattern that precisely matched the rodent's zigzagging path.

In our brains, the hippocampus is a seahorse-shaped structure buried deep within the medial temporal lobe. A bird's hippocampus sits on top of the brain like a button or a little toadstool. But in both bird and boy, this whiff of tissue harbors our mental maps—and our memories. In fact, our recollections appear to be all bound up with where we experienced an event. New research shows that when we recall an event, the place cells in our hippocampus that store the location of that event fire again, helping us to locate a memory in both space and time. This explains why retracing your steps can help you remember what you were looking for. Your memory of a thought is married to the place in which it first occurred to you.

In birds, the hippocampus plays a critical role in processing spatial information. A bigger hippocampus generally means better spatial ability. Families of food-caching birds have a hippocampus more than twice the size expected for birds of their brain size and body weight. For instance, in relative terms, a chickadee hippocampus is two times bigger than a sparrow's.

Hummingbirds can really crow in this respect. Relative to their whole brain size, they have a bigger hippocampus than any other bird, two to five times larger than that of caching and noncaching songbirds, seabirds, and woodpeckers. One big hummingbird, known as the long-tailed hermit, has a brain only as big as that of an American redstart but boasts a hippocampus almost ten times larger—the better to remember the location, distribution, and nectar content of the gingers and passionflowers it feeds on in Venezuela and Brazil.

Brood parasites such as honeyguides and cowbirds also have a big hippocampus compared with nonparasitic birds in the same family. "This makes sense," says Louis Lefebvre. "A honeyguide has to find a suitable nest to drop her eggs into at just the right time. If she puts them in a nest where chicks will be hatching the next day, her babies will be bumped off as runts; if she drops it in too early, the host bird may not be ready to lay or incubate. So she has to monitor the position of nests and the stages they're in."

Female cowbirds have a larger hippocampus than males—and, as Melanie Guigueno and her colleagues at the University of Western Ontario recently found, they also have more spatial prowess. In most animals, it's the males who have superior spatial abilities, but in birds, brood parasites upend this stereotype. Only female cowbirds locate, monitor, and revisit the nests they parasitize. They search the canopy and watch nest-building activity to spot host nests. Then, before sunrise, they locate the nests in the dark and lay their eggs there. In a laboratory study, Guigueno found that female cowbirds were far more adept at spatial memory tasks than their male counterparts. This suggests that superior spatial ability is not inherently male, but evolves in relation to the ecological demands of the brood-parasitic way of reproducing.

Homing pigeons have a heftier hippocampus than other pigeon strains bred for their fancy features, such as fantails, pouters, and strassers. But this hippocampal prowess is not genetic. It's hard earned.

Not long ago, a clever experiment revealed that the size of a homing pigeon's hippocampus depends upon its use. Scientists raised twenty homing pigeons in the same loft near Düsseldorf, Germany. After the birds fledged, half were allowed to fly around and learn the location of their loft and its surroundings. They also participated in several races of distances up to 175 miles. The other ten birds were confined to a loft spacious enough for the birds to fly around in freely, so they got roughly the same amount of physical activity as their more liberated counterparts, but not the navigational practice. When all the birds had reached sexual maturity, the scientists measured the volume of their brains and their hippocampi. Those pigeons with experience navigating had a hippocampus more than 10 percent larger than the inexperienced birds. It's not clear what biological mechanism is responsible for the ballooning, say the scientists. "Existing cells could increase their cell body size," they speculate, or new supporting brain cells could be added (though probably not neurons), "or there could be increased vascularization."

IN ANY CASE, the size of a pigeon's hippocampus may reflect experience and how often its navigational skills are called up. In other words, it may be shaped by use. British researchers discovered that this appears to be true for humans, too, in a now-famous study of some highly skilled modern navigators, London cabbies. Before aspiring cabdrivers can get an operating license in London, they must pass a stringent exam known as the Knowledge. This involves memorizing the spatial layout of some twenty-five thousand streets, as well as thousands of landmarks, in what has been deemed by popular survey "the world's most confusing city." Mastering knowledge of London's byzantine byways takes two to four years. The scientists found that cabbies who have been driving for a number of years have more gray matter in the rear portion of their hippocampus compared with cabbies who are still green or London bus drivers.

This raises a troubling question. If our human navigational efforts shape our hippocampus, what happens when we stop using it for this purpose—when we lean too hard on technology such as GPS, which makes navigation a brain-free endeavor? GPS replaces navigational demands with a very pure form of stimulus-response behavior (turn left, turn right). Some scientists fear that overdependence on this technology will shrink our hippocampus. Indeed, when researchers at McGill University scanned the brains of older adults who used GPS and those who didn't, they found that the people accustomed to navigating on their own had more gray matter in the hippocampus and showed less overall cognitive impairment than those who relied on GPS. As we lose the habit of forming cognitive maps, we may be losing gray matter (and along with it, if Tolman is right, our capacity for social understanding).

WE KNOW WHERE a bird's mental map likely resides. But how big can it be?

I'm thinking about this one early October morning on the beach at Cape Henlopen, Delaware. The day has dawned cold. Water temperatures are plummeting. On the bay side of the cape, I'm hoping to glimpse an osprey. But the big birds are mostly gone, headed south to Peru or Venezuela to winter in the warm swamps of the Amazon.

However, it's still peak migratory season for some raptors, along with the songbirds they feed on. Across the Delaware Bay at Cape May, merlins are moving through, along with kestrels, peregrine falcons, sharp-shinned hawks, and Cooper's hawks, pausing to pluck small birds from their journeys to sustain themselves during their own travels. The little brown jobs are at Cape May in abundance. In the brushy areas of Hidden Valley and the farmland known as the Beanery are a glister of goldfinches and yellow-rumped and palm warblers, along with some late parulas, blackpolls, and red-eyed vireos.

A cold front can push tens or even hundreds of thousands of migratory songbirds through this area at one time, a wondrous sight if you happen to be watching from the dike at Higbee Beach. These neotropical migrants will rest and feed for a few days and then launch back into the night. I love to imagine autumn's night sky sprinkled with dark southbound birds.

As I round Henlopen to the ocean side of the cape, a thick bank of fog hovers off the coast. I watch it with curiosity for a moment as it rolls closer, like a giant gray wave. Suddenly it enfolds me in a blanket of salt damp. The dunes along the shore melt into mist, and I can't see three feet in front of me. It's strangely disorienting—but that's all; I can easily trace the shoreline and find my way back through the dunes.

It's a different matter being in fog at sea. John Huth, a physics professor at Harvard University, tells the story of setting off in his kayak into Nantucket Sound on a clear day at this very time of year. Without warning, he became engulfed in dense fog. A veteran kayaker, Huth had been careful to note important cues before he left, especially the direction of wind and waves. "I stayed close to the shore," he writes, "and anytime the fog obliterated landmarks, I knew how to turn toward land." Two other kayakers paddling nearby that day were not so fortunate; they apparently lost their bearings, were swamped by heavy waves, and drowned.

As Huth points out, early human navigators could read natural cues to find their way. Polynesian voyagers created a natural compass by remembering the positions of rising and setting stars. Arab traders used the smell and feel of winds to traverse the Indian Ocean. The Vikings used the position of the sun to determine time and orientation. Navigators of the Pacific islands read the waves. With learning, we can find our way through close observations of sun, moon, and stars; tides and currents; wind and weather. (I was interested to learn that roughly a third of the world's languages describe the space occupied by one's body not in terms of right and left but with cardinal directions. Those who speak such languages are more skilled at staying oriented and keeping track of where they are, even in unfamiliar places.) But without a map or GPS on hand, most modern humans are hopeless at navigating.

Birds migrating in the ocean of air, on the other hand, rarely lose their way, even in darkness or fog. Like pigeons, they rely on available compass cues from visual landmarks, the sun, and magnetic fields.

At night, some use the stars, but not in the way you might think. They carry no map of star patterns but learn the apparent rotation of the night sky around the North Star. In the first summer of their lives, fledglings search the starry night sky for its center of rotation. In the northern hemisphere, this rotational center is the North Star, which birds learn to interpret as north. They orient themselves away from the star to head south. Once their stellar compass is fully established (which takes only about two weeks), birds can orient by the stars even if only some are visible.

I know that navigating by celestial cues is not necessarily a sign of high intellect. After all, dung beetles—best known for sculpting little balls of animal feces that they later eat—use light from the Milky Way to orient themselves at night. Still, it seems a marvel to me that birds can glean a north-south orientation by learning the rotational patterns of stars.

Migrating birds do occasionally get thrown off course hundreds or even thousands of miles by natural events such as storms. These are, in a way, natural displacement experiments on a grand scale. The ability of most migrants to find their way back to their target goal after such disorienting dislocations suggests that their mental maps may be very big indeed.

I HAD HOPED TO visit Cape Henlopen a year earlier, but my plans were foiled by Hurricane Sandy. Just a day or two before my scheduled arrival that year, the superstorm was barreling up from the south, its eye aimed straight at the cape. I'm glad I didn't brave it. The hurricane hit Henlopen dead-on, swamping roads and destroying bridges, piling sand into parking lots and side streets.

After Sandy passed, the whole eastern edge of the continent was swarming with vagrants. It's an interesting term, commonly used for someone who travels idly with no means of support. The word comes from the Latin root *vagari*, meaning "to wander." A vagrant bird that has wandered or been blown off course is by its very nature unusual and draws a twitch of birders keen on spotting that odd bird out of place, a quick addition to the life list.

In the wake of Sandy, birdwatchers at Cape May reported seeing more than a hundred pomarine jaegers—predatory seabirds that were likely blown inland while migrating south from their nesting grounds in the Arctic to winter in tropical seas. Hundreds more were spotted as far inland as Pennsylvania, flying south along the Susquehanna River. Sooty terns, red phalaropes, a Sabine's gull, a Cory's shearwater, and a tropicbird all ended up in Manhattan. A scattering of northern lapwings—European shorebirds—turned up in the open fields along New England coasts. And a Trindade petrel, a deepwater bird that normally spends its days over the open Atlantic Ocean off Brazil, dropped down near Altoona, Pennsylvania, west of the Appalachian Mountains and two hundred miles from the coast. It wasn't there for long, though. After the wind died down, the bird headed south.

If you want to add an accidental to your life list, you have to act quickly. Usually they're on their way within a day, apparently knowing just which way to go.

THAT EXPERIMENTAL SHUTTLING of white-crowned sparrows from the Pacific Northwest to Princeton, New Jersey, was a more extreme version of Hurricane Sandy, a deliberately massive displacement experiment. The scientists who conducted it were hoping to shed light on the size of a bird's navigational map—and they did.

That the sparrows (even those with minimal experience) could so quickly adjust and correct their course after a three-thousand-mile displacement suggested the existence of a vast mental navigational map encompassing at least the continental United States and possibly the globe.

The experiment also suggested that the map is experience based. The young, completely inexperienced birds in the experiment did not fare so well. They failed to find their way back across the country and instead, guided by instinct alone, just flew south.

Birds are not born with their maps. They learn them. Some do so by following the routes of the adult birds around them: whooping cranes, for instance. Inexperienced whoopers shadow adults along migration routes, which is why scientists can train naïve captive whoopers to follow a microlight aircraft as if it were an avian pied piper.

But trailing a parent is not always possible. A fledgling puffin, for example, leaves the isolated North Atlantic cliff slopes and islands of its birth at night, well before adults leave the colony for the winter. Likewise, a young cuckoo doing the English season in Norfolk can't follow his parents to the rainforests of the Congo because they have already gone south before he has fledged from the nest of his foster parents.

Still, a young migratory bird (provided it has not been chick-napped and shipped across the country) somehow manages to find its way hundreds or thousands of miles to its wintering grounds, though it has never been there before. To do so, it relies on a bit of wondrous genetic intelligence, an innate "clock-and-compass" program that tells it to fly for a certain number of days in a certain direction. The clock is an internal timekeeper under genetic control that dictates the number of flying days. We know this because a caged migratory bird will demonstrate a set amount of migratory restlessness, the *Zugunruhe*, which is tightly correlated with the distance it usually migrates. As for the compass piece: At least some juvenile birds carry an inherited one-direction compass that's specific to their species and sets them on the proper course. To stay on that course, they rely on the compass cues adults use, including sun, stars, the geomagnetic field, and the polarized light cues available at sunset. (Twilight is a rich source of information for navigating animals of all types. It's the only period in the day when birds and other animals can combine light-polarization patterns, stars, and magnetic cues.)

It's hard to imagine how this innate program works, especially for birds with extremely precise and complex routes. But somehow, coded in their genes and passed from one generation to the next is species-specific information on both direction and distance.

For the return trip and later migrations, birds no longer depend on inherited information. As they travel, they build a cognitive map that allows them to use true navigation to find breeding or wintering grounds they've visited before—and even to correct for displacements by wind, storm, and other natural events. For some birds, at least, this map in the head would appear to be immense, spanning continents and even oceans. Witness the white-crowned sparrow or the Manx shearwater. In one displacement experiment with shearwaters, birds transported thirty-two hundred miles to Boston from their nesting island in Wales found their way back home in just twelve and a half days.

WHAT IS THIS MAP made of? It may work like our Cartesian coordinate system, with different environmental cues that vary predictably along gradients providing information about latitude and longitude. To use these gradients, says Richard Holland of Queen's University, Belfast, a bird "would have to learn that they vary predictably in intensity with space (and possibly time) within their home range and extrapolate this beyond the learned area."

But what are the sensory cues contributing to the map's coordinates? Does it even have coordinates? Despite a welter of studies over the past four decades, we're still trying to uncrumple the much crumpled matter of mapping cues.

The gradient map may be partly geomagnetic. Holland and a colleague lately made a curious discovery. The pair picked up several European robins that had stopped over at a resting stage on their migratory path and exposed them to a strong magnetic pulse, temporarily disrupting their magnetic sense. Then they let the birds go. The young robins (which had no previous navigation experience) seemed unbothered by the pulse and continued on their expected path, guided by their innate program. But the adults flew off in the wrong direction. The researchers speculated that the adult birds had built up magnetic maps in their heads during their migrations, which they used to help them navigate on subsequent journeys. The pulses may have "reset" those maps, confusing the birds.

Another recent experiment, this one tinkering with Eurasian reed warblers, also points in this direction. A team led by Nikita Chernetsov and Henrik Mouritsen caught warblers on their migratory route north from Kaliningrad, Russia, on the Baltic Sea, to southern Scandinavia. In half the birds, the scientists cut the nerve that runs from beak to brain, the so-called trigeminal nerve, which is thought to transmit magnetic information to the brain. Then they displaced all the birds more than six hundred miles to the east of their normal migratory path. The warblers with an intact beak-to-brain trigeminal nerve quickly reoriented toward the northwest and their normal breeding grounds. But the birds with a severed nerve headed northeast as if they were still on their normal migratory path. The remarkable thing was that the birds knew where north was, but they had lost their ability to fix their position. In other words, they appeared to have lost their map sense.

We humans are highly visual creatures, especially when it comes to spatial matters. It's hard to wrap our minds around a map made of cues we can't see.

Here's another one. According to Jon Hagstrum, a geophysicist at the U.S. Geological Survey who has studied bird navigation for more than a decade, natural infrasonic signals, low-frequency noises in the atmosphere beneath our range of hearing but perhaps audible to birds, may be part of a map that helps them find their way.

It may also cue them to the arrival of storms. A startling example of the apparent ability of some birds to anticipate impending storms recently came to light by accident. It was April 2014, and researchers at the University of California, Berkeley, were testing whether a population of tiny golden-winged warblers breeding in the Cumberland Mountains of eastern Tennessee could carry geolocators on their backs. The birds had arrived only in the past day or two after a 3,000-mile journey north from their wintering grounds in Colombia. The team had just attached the gizmos to the tiny warblers when all the birds suddenly flew the coop, spontaneously evacuating their nesting grounds. The scientists later learned that a huge "supercell" spring storm was headed their way, one that would spawn eighty-four tornadoes and kill thirty-five people. The warblers left twenty-four hours before the devastating storm hit and flew in all directions, some as far south as Cuba. After the storm passed, they flew straight back to their nesting site—for some, a round-trip of almost 1,000 miles. The scientists conducting the study suggest that the birds may have been warned by the deep rumble of the superstorm when it was still 250 to 500 miles away, picking up on the strong low-frequency infrasounds generated by such tornadic storms. These can travel for hundreds to thousands of miles but are inaudible to humans.

Infrasounds are produced by many natural sources, but primarily by oceans. Interacting waves in the deep ocean and movement of sea surface water create a kind of background noise in the atmosphere that can be detected anywhere on earth with the help of a low-frequency microphone. In addition, pressure changes on the seafloor generate seismic waves in the solid earth that can interact with the atmosphere at the ground surface—"like a giant speaker cone," says Jon Hagstrum—to produce infrasound waves that are radiated outward by hillsides, cliffs, and other steep terrain and can travel great distances. Each location on earth possesses a kind of sound signature, then, shaped by topography. In Hagstrum's view, birds may use these sound signatures to navigate and to locate their lofts "infrasonically."

"Similar to the way we see a landscape, I think birds are hearing it," says Hagstrum. "Farther away, they are probably listening to the sounds produced by larger landscape features, and as they get closer the features get smaller." In other words, a pigeon may know just what the area around her loft "sounds" like. "Pigeons with frosted lenses covering their eyes can return to within a kilometer or two of their loft but need to see it for their final approach," says Hagstrum. "I think this is the smallest possible area that can produce infrasounds loud enough for a pigeon to hear."

There are plenty of skeptics. "The anecdotal evidence is certainly fascinating," says Henrik Mouritsen, "but the key question that needs to be answered by anybody suggesting infrasound as a bird orientation cue is sensory: Can birds sense infrasound at all? There's no evidence of this. And secondly, can they determine the direction it's coming from? This normally requires a very large distance between the ears (as in elephants and whales)," suggests Mouritsen. In his view, a much more likely explanation for the ability of those warblers in Tennessee to detect that distant superstorm is not infrasound but changes in atmospheric pressure, which birds are known to be able to sense.

However, if Hagstrum's infrasound theory is true, it could shed light on the disappearance of Whitetail and those sixty thousand other pigeons that vanished from England and France almost two decades ago. Intrigued by the disappearance of so many pigeons in that disastrous race, Hagstrum dug around the historical records to see if any unusual sound events might have coincided with the race. Sure enough, there was a big one that day: Just as the racing pigeons were heading out across the English Channel, their flight route crossed paths with a Concorde SST just leaving Paris. When the plane went supersonic, says Hagstrum, it laid down a "sonic boom carpet" so loud it obliterated the pigeons' navigational acoustic map, completely disorienting them.

Hagstrum's idea may also help to explain certain Bermuda Triangles of pigeon homing—places where pigeons tend to disappear or get hopelessly lost. The geometry of terrain in these spots may create what he calls "sound shadows," which disrupt their acoustic bearings.

The idea remains highly controversial. Says Richard Holland, "these correlations are compelling, but they're just that, correlations"—in the case of the pigeon race, a correlation between infrasonic disturbance (the sonic boom) and orientation disturbance (the missing birds). "This is weak evidence," says Holland. "No experiment has yet demonstrated any effects of infrasound on bird navigation."

ODOR MAY PLAY INTO the map, too—another concept that stretches the human imagination and inspires debate, though this theory is backed by substantial experimental evidence. The idea that odor cues might factor into bird navigation began more than four decades ago when Floriano Papi conducted an experiment on pigeons in Tuscany. The Italian zoologist and his colleagues severed the olfactory nerves in a group of pigeons and released them at an unfamiliar site. The birds never returned, while their intact companions quickly flew back to the loft. At around the same time, German ornithologist Hans Wallraff found that pigeons at their home loft sheltered from the wind by glass screens were unable to find their way home. Thus was born the olfactory navigation hypothesis, which suggests that pigeons learn to associate the wind-borne smells of their home loft with wind direction and use this information to determine their way home.

The possibility that birds navigate with the help of an odor map may shed light on a weird evolutionary paradox that scientists have puzzled over for more than a decade. It concerns a fluke in the geometry of animal brains. If you look at vertebrate brains across different orders, classes, families, and species, a neat pattern emerges, a kind of universal scaling law. In nearly all vertebrates, components of the brain, from the cerebellum to the medulla to the forebrain, scale up predictably with the size of the brain as a whole. More often than not, you can predict the size of the brain component from the total brain size. Structures in the brain that evolved more recently are usually bigger.

Nature sometimes yields such lovely rules of thumb.

However, "the 'late equals large' principle has one important exception," says Lucia Jacobs, a psychologist at the University of California, Berkeley: the olfactory bulb. It's a renegade in nearly every way.

The olfactory bulb is an ancient part of the brain devoted to the sense of smell and found universally in vertebrates. It's often smaller than expected vis-à-vis the rest of the brain, or larger. (The largeness is especially odd, given its old evolutionary age.) And it varies in size across animals in the same order, the same class, or the same family. This is true for birds. Petrels and other seabirds such as shearwaters and albatrosses have bulbs roughly three times the size of those in songbirds. In the American crow, the bulb is only 5 percent of the length of the bird's cerebral hemisphere, whereas in the snow petrel, it's more than 35 percent.

The largeness of the bulb in some birds has presented a conundrum. In the brain, big usually means important. It's called the principle of "proper mass"—the more brain space devoted to a function, the greater its importance to the biology of the animal. But for a long time, scientists thought birds couldn't smell much. They didn't display any of the more obvious nose-inspired behaviors—sniffing butts or snuffling truffles. Birds were more like us, it seemed, eye-minded creatures with highly evolved and sophisticated visual systems. "An extraordinary development of one set of organs is never accomplished but at the expense of some other set," wrote one ornithologist in 1892. "In this case the organs of the sense of smell have been the martyrs."

That view has radically changed. The shift began in the 1960s with experiments revealing that pigeons exposed to a stream of scented air showed a spike in heart rate. For their hearts to react this way, they had to be smelling something. Later, scientists planted electrodes in the birds' olfactory bulbs. To their surprise, they found the same pattern of cell firing in response to odor stimulation that occurs in the olfactory bulbs and nerves of mammals.

Since then, nearly every species tested has demonstrated some olfactory talent, from kakapos and starlings to ducks and small petrels known as prions. Kiwis, flightless nocturnal birds of New Zealand, find their invertebrate feasts by tracking their scent through nostrils on their long beaks. Vultures are able to trace the odor of a decaying animal carcass miles away and approach it by flying upwind. Blue petrels—seabirds that forage over featureless waters for scattered krill, fish, and squid—can detect the odors of their prey even before they fledge, at minute concentrations. These petrels nest in dark burrows, and on moonless nights they appear to rely on scent to find their way back through dense colonies to their individual home burrows.

All of these more obviously scent-oriented birds have big olfactory bulbs. But even species with much smaller bulbs, such as songbirds, appear to pluck odors from air, soil, and vegetation and use them to detect predators or plants protective against harmful microbes. Blue tits feeding their nestlings won't enter a nest box if it has been laced with the scent of a weasel. And they'll sniff out fresh yarrow, apple mint, and lavender, and ferry fragments to their nest cups to protect their chicks from pathogenic bacteria and parasites. Small seabirds called crested auklets don't let their modestly sized olfactory bulbs prevent them from plunging into an odorous social rite every summer, burying their noses in the napes of other auklets to scope out their scent—an aroma reported to smell like freshly peeled tangerines, perceptible only in the breeding season but strong enough to be detectable even to human noses as far as a half mile downwind. Zebra finches, whose bulbs are tiny indeed, use their sense of smell to spot their relatives, just as mammals do, to avoid inbreeding and facilitate cooperation with their kin.

But why the wide variation in bulb size? Do the discrepancies just reflect differing demands for acute smell called for by different foraging pressures or social lifestyles?

Lucia Jacobs has another explanation. A specialist in cognition and brain evolution, Jacobs proposes that the olfactory bulb first evolved in all vertebrates, including birds, not for hunting or foraging or ducking predators or communicating or finding mates—but rather, she says, for "animals to decode and map patterns of odorants for the purpose of spatial navigation." The universe of smells is super dynamic, with constantly moving cues. "It demands a neural architecture adapted to learning complex patterns," explains Jacobs. In fact, this could have been the primary driving force in the evolution of associative learning, Jacobs suggests, the ability to learn and remember the relationship between unrelated items—for example, the scent of a certain mineral or tree and the direction toward home. The size of the bulb in birds today correlates more tightly with their ability to navigate by olfactory cues than with their capacity to discriminate odors for food or protection against predators. The homing pigeon, for instance, has an impressively big olfactory bulb compared with its nonhoming domestic pigeon cousins, which otherwise share its lifestyle.

CERTAIN BIRDS WITH big olfactory bulbs do seem to possess some sort of detailed smell map. Anna Gagliardo of the University of Pisa has found that Cory's shearwaters, pelagic birds of the Atlantic Ocean, appear to use odor maps to find their way around the sea. Shearwaters wander the wide oceans searching for food, but each year they manage to find the same tiny island on which to breed and raise their young. To find out how they do it, Gagliardo and her colleagues removed two dozen shearwaters from their nests on the Azores during nesting season and loaded them onto a cargo ship headed to Lisbon. Some of the birds were fitted with small magnetic bars that scrambled their magnetic sense; others had their nostrils washed with zinc sulphate, temporarily obliterating their sense of smell. Once the ship was hundreds of miles from the breeding island, the birds were released. Those with the mixed-up magnetics found their way back, but those with the neutralized noses were completely confused and meandered around the ocean for weeks. Some never returned to their islands.

A NAVIGATIONAL MAP of smells would not look like any bicoordinate map we've ever seen. Jacobs imagines a dual mapping system for olfactory space, based on the work of Papi, Wallraff, and other researchers. The first part is a low-resolution map made of various odor plumes that mix at different gradients to create a grid dividing olfactory space into subregions she calls "neighborhoods." These odor plumes may consist of different ratios of so-called volatile organic compounds, chemicals in the atmosphere that may be the source of odors. When Wallraff sampled air at ninety-six different sites within a 125-mile radius of a pigeon loft in southern Germany, he found that these ratios increased or decreased along fairly stable spatial gradients. For a pigeon, shifts in the ratios may translate into shifts in scent. In other words, different areas have different smells.

Think of a pigeon in its home loft. There's a scent of lemon trees in one direction and the aroma of olive trees in the other. If the bird flies toward the lemon trees, the odor of lemons will grow stronger while the smell of olives will grow fainter. Drop a pigeon in a "neighborhood" somewhere in between (of, say, 20 percent lemon, 80 percent olive), and from the particular mix of gradients it will glean information about the direction of home.

The second part of the map is a collection of odorous landmarks—blends of odors unique or specific to a particular locale. Imagine an aromatic rendition of the Statue of Liberty or the Tower of London.

This notion of an odor map is still a topic of hot debate, and there are significant problems with it. Odors are airborne and move with the winds. So it seems unlikely that smells would congeal to form any sort of stable bicoordinate map. "Obviously, the question of turbulence looms large," says Jacobs. But birds and other animals are pretty good at decoding turbulence, she says. And as it turns out, the distribution of at least some odors in the atmosphere *is* fairly stable, creating predictable spatial gradients that might be useful for a bird navigating over distances of hundreds of miles—though probably not beyond that.

To complicate matters, there's a possibility that odors may act more as motivational cues than as navigational ones. One study found that in young pigeons, odors appear to activate other navigational processes. If this study holds true, says Richard Holland, smelling "non-home" odors "may trigger a bird to access a navigation system based on other cues."

Still, a recent experiment by Holland and his colleagues showed that adult catbirds deprived of their sense of smell and then displaced from Illinois to Princeton, New Jersey, were unable to correct for the displacement the way their counterparts with active noses could. Moreover, when scientists looked inside the brains of migratory birds during their *Zugunruhe*, they saw activity in both the visual and olfactory areas of the brain, suggesting that smell does indeed play a part in migratory behavior. It's just not clear what that role is.

It's an intriguing idea: a mental map made up—at least in part—of mosaics of odors and curling signposts of smell. Jacobs believes birds may use the neighborhood system as a rough map to sort out their general position and determine flying direction. The landmark system would take time to learn but would ultimately yield a map of higher spatial resolution. In this scenario, then, olfaction may supply two kinds of mapping cues. Over evolutionary time, Jacobs suggests, the hippocampus became specialized to process and integrate these two olfactory information streams. Eventually, it "learned" to integrate other kinds of sensory cues, such as magnetic signals and sound. This might explain why the olfactory bulb is such an outlier when it comes to brain scaling. With the evolutionary shift in some species to the use of other sensory information for navigation, the olfactory bulb dwindled in size.

I FIND IT ODDLY thrilling that the mental maps of birds remain . . . well . . . unmapped. There's no clear evidence that any one sensory cue is the whole story. Which cues a particular bird uses on any particular journey may depend on the scale of the trip, or what's handy, or the environmental conditions (like a kayaker in fog, it may resort to lesser ones when the chief ones aren't at hand), or simply its own individual predilections.

For instance, which cues a pigeon uses to home in on its loft may depend on its life experience and its own quirky choosing. In Blaser's study of homing pigeons, she found that the pigeons never beelined to their target; each time, they took a slightly different route—"a compromise between their chosen compass direction, topographic factors, and their own individual flight strategies," she says. A lot depends on how a pigeon grows up—and where. A pigeon raised in a loft without ambient odors orients using other cues and isn't affected when deprived of its sense of smell, according to Charles Walcott. Likewise, sibling pigeons raised in different lofts have different responses to magnetic anomalies: One finds its way despite the weird magnetic pattern; the other is bewildered by it and loses its sense of direction.

Individual birds are also just eccentric and seem to use their orientation cues according to their own styles. Walcott tells the story of one pigeon raised near a prominent hill in Massachusetts. When released at an unfamiliar site, he always flew to the nearest mountain before flying home—unlike any of the other birds raised in his loft. Another pigeon was a champion distance navigator, but once he got within six miles of his loft, says Walcott, he just kind of gave up and landed in a garden somewhere. In this arena, as in all aspects of bird (and human) life, idiosyncrasy and opportunism may prevail.

Like an executive who enjoys having two cell phones and a laptop tuned to the Weather Channel, a pigeon may rely on all kinds of available information for guidance. She may use multiple and redundant cues to find her way, and mental maps unlike anything we've ever encountered. Her spatial grid may not be bicoordinate at all, but multicoordinate, layered with a still mysterious mix of sun, star, and geomagnetic cues, sound waves, and swirling signboards of smell, all thoroughly integrated.

THIS NOTION SEEMS TO dovetail (if you'll excuse the expression) with a new theory about the overall organization of bird brains—and our own.

In the lingo of neuroscience, brains are known as "massively parallel, distributed control systems." Roughly speaking, that means that they contain a colossal number of little "processors"—neurons—that operate in parallel but are distributed all over. The problem in a brain, then, is how to pull together all those distributed resources—the totality of what an animal knows—to meet a challenge (such as navigation) or respond to unpredictable circumstances (like a storm).

It's called cognitive integration. The brain of a bee, with only a million neurons, does it. So does the human brain, with its 100 billion neurons.

"Humans excel at cognitive integration," says Murray Shanahan, a computational neuroscientist at Imperial College London—although he admits that failures are commonplace, "as when I remove the U-bend of my sink," and then "pour its dirty contents back into the plughole, causing an unwelcome deluge." (Or the equivalent sink lore of my family: When, minutes before our big annual Christmas party, my mother stood at the sink in dismay, having just poured a cauldron of mulled wine for fifty through a mesh strainer directly into the drain, leaving only a clump of damp cloves, peppercorns, and bay leaves to serve to her guests.)

True navigation is a triumph of cognitive integration, says Shanahan.

To achieve it, a certain pattern of connectivity in the brain is required. Information about landmarks, distances, spatial relations, memories, sights, sounds, and smells must all funnel into a core of important brain regions and then fan out between them. This, he explains, "results in an integrated response to the [bird's] ongoing situation."

To figure out how this connectivity might work in a typical bird brain, Shanahan co-opted a team of neuroanatomists to analyze anatomical studies of pigeon brains. (Pigeons were a good species to pick, he says, because they can perform noteworthy feats of cognition.) Drawing on more than forty years' worth of studies tracing pathways between brain regions in pigeons, the team created the first large-scale map, or wiring diagram, of the pigeon brain, showing how different regions in a typical bird brain are connected to process information.

The surprise?

The map the team came up with looks a lot like such connectivity maps for mammals, including humans. Though birds have radically different brain architecture from ours, when it comes to connectivity, their brains seem to be organized in a similar fashion. Shanahan sees in the similarity what he calls a common blueprint for high-level cognition. In simplified terms: The human brain is thought to be a so-called small-world network, not unlike Facebook. Different modules—or regions—of the brain are connected by a relatively small number of neurons known as hub nodes. These hub nodes connect with lots of other neurons, sometimes over long distances, to provide a short connective link between any two nodes in the network. (Think of someone with thousands of "friends" on Facebook.) The hub nodes linking parts of the brain important in cognition—such as long-term memory, spatial orientation, problem solving—together make up the brain's "connective core."

Shanahan found that the hub nodes in the pigeon's hippocampus, in particular—so crucial to navigation—had very dense connections to other parts of the bird's brain.

The idea is this: If a migrating lapwing or reed warbler is blown halfway across the country by a storm, perhaps the information her senses gather from all of her sources—from the scents of land and sea, from magnetic signatures and anomalies, from the slant of sunlight and the starry pattern of night skies—all funnels into the connective core in her brain, where it's integrated and then fans out to the brain regions that will help guide her to her natal ground.

In a bird brain, then, a small-world network may create a big-world map. So that a hummingbird can find her way to David White's feeder each spring. So that an Arctic tern can journey like a guided missile from one bright pole to the other. So that one cool April morning after being away for five years, Whitetail the racing pigeon can finally hightail it home.

eight

SPARROWVILLE

Adaptive Genius

It is not the strongest of the species that survives, nor the most intelligent . . . It is the one that is the most adaptable to change." These words are often attributed to Charles Darwin (and, to the embarrassment of the California Academy of Sciences, once etched as such into its stone floor), but they're actually from the pen of the late Leon Megginson, a professor of marketing at Louisiana State University.

I'm reminded of the good professor's words early one May morning. A group of us has gathered for a spring bird count at the Crossroads Shopping Center in Albemarle County, Virginia. Our first birds: a common grackle, a house finch, and a family of house sparrows nesting above a sign that says MOM'S LAUNDROMAT.

"We call them 'parking lot' birds," says my birding friend David White.

Where can you find a nesting sparrow? In the rafters of buildings and the clips holding gutter spouts to houses. In the ventilator holes beneath flat roofs, inside streetlights, in flowerpots on your porch. Rarely far from any man-made structure. One family of sparrows nested for generations in a coal mine hundreds of feet underground, kept alive by the offerings brought by miners. I once found a sparrow's nest in the tailpipe of an abandoned Toyota sedan.

"What did these birds do before civilization?" asks David.

Passer domesticus. As its Latin name suggests, the house sparrow is the migrant's polar opposite. Like a pushy houseguest, it's invited in, but often overstays its welcome. It's a permanent resident through most of its range—and remarkably sedentary, clinging to the locality of its chosen home, foraging close to roosting sites, breeding close to its natal colony. And yet the house sparrow's rapid spread around the globe is legendary.

In his book *Biology of the Ubiquitous House Sparrow*, Ted Anderson cites a theory of the sparrow's origin that is revealing of its nature. The theory proposes that the bird was always "an obligate commensal of sedentary humans." It came into its own as a species only since the advent of agriculture in the Middle East, approximately ten thousand years ago. Other theories place its origin earlier than that, more like half a million years ago, based on fossil evidence found in a cave near Bethlehem in Palestine. In any case, so highly skilled has the house sparrow become at adapting to any environment occupied by humans that it has been called the ultimate opportunist, our avian shadow.

DOES THE SPARROW'S KNACK for adapting to human habitats call for any special kind of smarts? What about birds that don't possess it?

These are not casual questions. Birds are facing change on a scale unknown in their evolutionary history. This is a result of the Anthropocene— the new epoch of man-made change that is contributing to what has been called the sixth mass extinction. The habitats that birds have occupied for millions of years are going the way of croplands and cities and sprawling suburbs. Exotic species are displacing natives. Climate change is shifting the bands of rainfall and temperature that birds count on for feeding, migrating, and breeding. Many species don't tolerate these changes well. Some do.

Is there something special in the cognitive toolkits of house sparrows and their like—pigeons, turtledoves, and other so-called synanthropes drawn to settle near humans—a set of mental skills that allows them to thrive in a place no matter how altered or degraded it may be?

Or perhaps it's the other way around. Maybe the changes we're wreaking are changing the birds themselves, shaping the nature of their brains and behavior. Are we humans selecting for a certain breed of bird intelligence? A sparrowlike smartness?

THE ORNITHOLOGIST PETE DUNN calls *Passer domesticus* the "Sidewalk Sparrow." Before 1850, there were no house sparrows in North America. Today there are millions. You have to hand it to them. The first sixteen birds said to have been introduced to Brooklyn in 1851 to control a plague of moths may not have taken immediately to the New World, but another bigger shipment imported from England the following year did, and in a big way. The birds got some help from individuals and naturalization societies bent on populating their gardens and parks with plants and animals from the Old World, which no doubt accelerated their expansion. But even so, the success of their spread is staggering.

The transplants found a land much to their liking, rich in grain and horse droppings. They multiplied and dispersed rapidly, spilling into farming districts, where they exploited every food source they could find—grains, small fruits, and succulent garden plants, such as young peas, turnips, cabbage, apples, peaches, plums, pears, and strawberries. Soon they were considered a serious pest. In 1889, just a few decades after the house sparrow's introduction, sparrow clubs were formed with the sole objective of destroying the birds, and county and state officials were offering two cents a head for each sparrow killed.

Before long, the birds had spread across the United States and Canada, adapting to environments as extreme as Death Valley, California, at 280 feet below sea level, and the Colorado Rockies at more than 10,000 feet above sea level. They moved southward into Mexico, through Central and South America as far as Tierra del Fuego, and along the Trans-Amazonian Highway deep into the rainforests of Brazil. In Europe,

Africa, and Asia, they dispersed to northern Finland, the Arctic, South Africa, and clear across Siberia.

Now the humble house sparrow is the world's most widely distributed wild bird, with a global breeding population of some 540 million. It's found on every continent except Antarctica and on islands everywhere, from Cuba and the West Indies to the Hawaiian Islands, the Azores, Cape Verde, and even New Caledonia. Ted Anderson writes that he can sit in his living room listening to news reports from just about anywhere in the world on the radio or TV and hear the characteristic chirrup of house sparrows.

WHEN I WAS GROWING UP in Maryland, the house sparrow was slammed as a "bad" bird. Not just pesky, belligerent, and meddlesome, but thuggish, known for molesting and displacing "good" birds—martins, robins, wrens, and, especially, bluebirds.

It deserved the reputation. When one scientist, Patricia Gowaty, monitored bluebird boxes in South Carolina for six years in the late 1970s and early 1980s, she found twenty-eight dead adult bluebirds inside the boxes. Twenty of these showed violent trauma to the head or breast; "in 18 the crown was bloody, denuded of feathers, and the skull cracked," she writes. House sparrows were observed at eighteen of the twenty nest boxes on her visits before and after the deaths.

Circumstantial evidence to be sure. Gowaty never caught a house sparrow in the act of hammering the head of a bluebird. But still, on three occasions, she discovered sparrow nests built over the bodies of victims. The right wing of one dead bluebird, she wrote, "was stretched out and up and incorporated into the dome of the House Sparrow's nest!"

The house sparrow may be rightly branded as a thug or feathered rat, stigmatized as pernicious, even murderous. But whatever people say, the bird is a superb invader, skilled at ensconcing itself nearly everywhere it goes. Of thirty-nine known house sparrow introductions, thirty-three have been successful.

FOR THE PAST fifteen years or so, Daniel Sol has been thinking about what makes a bird like the sparrow slip so easily into anyplace it goes. Sol, an ecologist at Spain's Centre for Research on Ecology and Forestry Applications, calls it the invasion paradox: "Why can alien species succeed in environments to which they have had no opportunity to adapt and even become more abundant than many native species?" What gives some birds their edge in the face of radical change?

Imagine that one day dozens of different exotic bird species outside their natural range escape from their cages. Sol can tell you which ones will likely be around in twenty years, squabbling around our park benches, squawking from colossal nests on telephone poles, gathering in massive flocks that darken the sky, displacing the native species. He bases his predictions on what he has gleaned from observing the common traits of invasive birds around the globe.

In the past, scientists studying the invasion success of birds have focused on the influence of their nesting habits, migration patterns, clutch size, and body mass. Some years ago, Sol and his colleague Louis Lefebvre decided to see if brain size and intelligence might have anything to do with it. First they looked at the records of invasive birds in and around New Zealand, a region that has suffered from the presence of exotics of all kinds. Of the thirty-nine bird species introduced to New Zealand, nineteen succeeded in overrunning it; the other twenty did not.

When the duo studied the characteristics of the nineteen introduced species that "took" and those that failed to establish themselves, two pronounced differences emerged. The more successful invaders had larger brains. They also had more innovative, flexible behavior of the kind Lefebvre documented in his avian IQ scale.

When Sol later looked at 428 bird species that invaded areas around the world, the pattern held. Successful colonizers were brainy and inventive. Well represented among the intruders were those kings of innovation, the corvids: the house crow in Africa, Singapore, and the Arabian

Peninsula; the jungle crow in Japan; the common raven in the American Southwest. All big brained—and all considered pests in the regions they had invaded.

Successful amphibian and reptile colonists also have bigger brains than their less successful peers, as do mammals, including Homo sapiens, the so-called colonizing ape, invader of almost every land habitat on earth.

Big brains are costly in terms of development and maintenance. But they're thought to enhance a bird's survival by allowing it to rapidly adjust to unusual, novel, or complex ecological challenges such as finding new food or avoiding unfamiliar predators. It's called the cognitive buffer hypothesis. A big brain "buffers" an animal from environmental change by allowing it to adapt to novel resources—to try new foods and explore new objects and situations that a more "programmed" species might avoid. In other words, to be flexible enough to do things differently. For a bird to succeed in a novel or changed environment, says Sol, it must have a knack for doing something new.

ORDINARILY THERE'S NOT MUCH nourishment for a bird around a parking lot or a skyscraper. In Normal, Illinois, however, two ecologists watched house sparrows working their way along a line of parked cars in a parking lot, gleaning insects trapped in the radiators. Sparrows have also been observed foraging for insects late at night in the floodlights around the observation floor of the Empire State Building, eighty stories up (which goes to show, the observer noted, that "Manhattan is not an ornithological desert").

These are just a couple of pages from the house sparrow's book of tricks. In his tally of inventive bird behaviors, Louis Lefebvre studied 808 species. Many birds had only one innovation to their name. The house sparrow had forty-four.

The sparrow is known for nesting in unusual places—rafters, gutters, roofs, soffits, attic vents, dryer vents, pipes, ductwork—you name it. A

Missouri biologist discovered a truly novel nesting site when he noticed sparrows carrying food to a working oil pump in McPherson, Kansas. In inspecting the pumps, he found they contained three nests, all with young. Two of the nests were in constant motion with each cycle of the pump, seesawing about two feet up and down every few seconds.

Moreover, house sparrows line their nests with some unusual materials—feathers plucked from living birds, for instance, sometimes hundreds of them. In one weeklong period in spring, an observer at Victoria University in Wellington, New Zealand, caught several sparrows in the act, collecting feathers directly from the rump area of an incubating adult pigeon, six or seven an hour. "Typically the sparrow would arrive on the ledge," he wrote, "hop up onto the back of the pigeon, pull out a single contour feather and then fly away."

In some cities, you can find smoked cigarette butts in sparrow nests, which effectively function as a parasite repellent. Butts from smoked cigarettes retain large amounts of nicotine and other toxic substances, including traces of pesticides that repel all kinds of harmful creepy crawlies—an apparently ingenious new use of materials.

When it comes to foraging, house sparrows are particularly exploratory and inventive. They will go wherever the food is, no matter how foreign the place or the meal. They'll eat plant matter—primarily seeds—but also flowers, buds, and leaves, as well as insects, spiders, lizards, geckos, even an occasional young house mouse, not to mention a broad array of human refuse. And their foraging techniques are equally unconventional. Sparrows have been seen methodically plucking insects from spiderwebs straddling the railings along the River Avon in England. On the Hawaiian island of Maui, they've mastered the knack of pilfering morsels from the balconies of breakfasting tourists in the massive beachfront hotels. Instead of patrolling the hundreds of balconies facing the sea, the sparrows cling to the concrete walls between them and wait until breakfast is served. This saves the energy that might be expended flying to and fro to see who's eating breakfast or hovering before a balcony waiting for the baked rolls to arrive.

But perhaps their most famous Promethean feat is one that defied a fancy human invention. Some years ago, a pair of biologists watched with surprise and delight as house sparrows in a New Zealand bus station repeatedly opened an automatic sliding door that led to a cafeteria. The birds flew slowly past the sensor or hovered in front of it or landed on top of it, leaning forward and bending their necks until their heads triggered the sensor. They did this sixteen times in forty-five minutes. The new automatic door had been installed only two months earlier, but the sparrows had easily conquered its workings. The top of the sensor was covered with bird droppings.

The trick popped up in other places around New Zealand. According to one account, a sparrow at the Dowse Art Museum in Lower Hutt, New Zealand, was seen opening a double set of automatic doors leading to the cafeteria. A few minutes later the sparrow activated both sensors to return outside. Staff members at the cafeteria were familiar with the bird (they named him Nigel), having watched him trigger the sensors many times over the previous nine months. Despite the presence of sparrows and automatic doors with the same system of sensors in many countries, the observers noted, they could find no reports of the deed anywhere but in New Zealand. "It seems that either foreign ornithologists have not reported sightings," they wrote, "or that some sparrows in New Zealand are smarter than those in other countries."

COMPARE ALL THIS WITH the ruddy turnstone, a small wading bird near the bottom of the innovative-behavior totem pole. In his book *The Wind Birds*, Peter Matthiessen describes an early experiment on the shorebird's behavior by eighteenth-century English naturalist Mark Catesby: "Catesby provided a ruddy turnstone with stones to turn, the better to observe the feeding trait that gives the bird its name. In a time when scientific experiments were less complex than they are today, the bird was furnished systematically with stones that had nothing beneath them, whereupon 'not finding under them the usual food, it died.'"

MOST VERTEBRATES ARE EITHER fearful of strange objects or indifferent to them. But newfangledness of most kinds doesn't seem to faze a house sparrow. When Lynn Martin of the University of South Florida, Tampa, tested the sparrow's tolerance for novel objects such as a rubber ball and a toy plastic lizard by placing them near seed-filled feeding cups, he discovered a surprise. The house sparrows were not only unperturbed by the strange objects, they actually seemed drawn to them—happier to approach the seed-filled dishes when ball or lizard was present. Martin noted that this was the first record of a novel object actually being attractive to a vertebrate (apart from man).

If you're going to invade a new place, a love of novelty helps.

So does a fondness for hanging out in groups.

Sparrows are gregarious. They don't like to eat alone. Or bathe alone. Or roost alone. They forage in flocks, calling in other birds to join them in feeding. They roost in congregations that vary in number from a few individuals to hundreds or, occasionally, even thousands.

Group living offers clear advantages for sparrows, as it does for other birds. One is predator protection (almost anything will eat a house sparrow; the more vigilant eyes, the better). Another is faster food finding. A bird arriving from a particular direction at a communal roost with a visibly full crop may point to profitable foraging areas as well as a quick travel route.

Moreover, as it happens, larger groups of sparrows appear to solve problems more quickly than do individuals or small groups—at least according to recent work by András Liker and Veronika Bokony of the University of Pannonia in Hungary. The pair found that squads of six birds easily and consistently beat out groups of two at opening a tricky seed container, a clear Plexiglas box with holes drilled through the top. Each of the holes was covered by a lid with a small black rubber knob glued to it. To get at the seed, the sparrows had to either pull open the lid or vigorously peck it to detach it. The gaggles of six outperformed the pairs in every respect. They opened four times more lids, were eleven times faster at solving the problem, and got their seed seven times sooner. All in all, the bigger groups were about ten times more successful than the pairs. The scientists chalked up the flocks' greater successes to the odds that they included birds with varying abilities, experiences, and temperaments: "Large groups succeed because they're more likely to contain a diverse range of individuals," the team writes, "some of whom will be very good at problem-solving."

Research on other bird species confirms this. Among Arabian babblers, for instance, "once one individual in the group learns a task, the rest learn it relatively quickly," says Amanda Ridley. "New skills are more likely to be acquired in larger groups."

This is true for humans, too. Studies show that small but varied groups of three to five people crack intellectual tasks faster than even the brightest individuals. Psychologist Steven Pinker goes so far as to argue that group living and the opportunity it offered our ancestors to learn from one another set the stage for the evolution of humanlike intelligence.

Invading birds are constantly encountering new and challenging situations that require novel solutions, which groups come up with more quickly than loners do. "For species such as sparrows that live in habitats being continuously changed by humans," say the Hungarian scientists, "two heads are definitely better than one."

THERE'S A COROLLARY TO THIS: One house sparrow head is not necessarily like another.

That animals are individuals may seem obvious to pet owners. But for a long time, variation among members of a single bird species was considered mere noise. Birds of a feather were expected to behave alike. "There is a great tendency to see an animal do just what it is supposed to do," warned the ornithologist Edmund Selous. But "uniformity of action is in proportion to paucity of observation. . . . The real naturalist should be a Boswell, and every creature should be, for him, a Dr. Johnson." Birds are individuals and respond individually to all sorts of situations—which cues they use to navigate, how they respond to oxytocin-like molecules, whether they seek extra-pair matings, how they react to novelty. Like us, they vary in character and behavior. I suspect these variable behaviors reside in what we call "mind." But they also show up in the body—in the way a particular bird responds to stress, for instance. A stressful stimulus that elicits a big response in one bird (say, fight or flight) may evoke a mere ruffled feather in another. For instance, John Cockrem of Massey University in New Zealand, who studies stress responses in little penguins and other birds, has found that individual birds diverge considerably in how they respond to environmental stressors.

Again, these differences may be important to sparrows in adapting to novel or unstable environments. In dealing with, say, a big dangerous place like a city, it pays to mix it up.

Lynn Martin has caught house sparrows in the act of infiltrating new territory, giving him a window on the traits that mark the plucky birds at the front line of an invasion. Martin, an ecological physiologist, studies the house sparrows now swarming over Kenya. The birds were first introduced to the coastal city of Mombasa in the 1950s, probably on ships coming from South Africa. Martin began studying them as a graduate student in 2002, when sparrows were rare in Kenya. Now they're common in cities all the way up to the Ugandan border. (Like Ted Anderson, Martin uses the sparrows' chirruping heard on the radio and TV to help monitor their spread in Kenya.) He and his colleagues consider the distance from Mombasa as a proxy for the age of the sparrow populations. They're looking at differences between the old populations at the original site of introduction and the new populations at the edge of the range expansion, in cities such as Nairobi, Nakuru, and Kakamega.

Birds farthest from Mombasa, at the cutting edge of invasion, have revved-up immune systems. They also release more stress hormones known as corticosterones after being caught. The scientists suggest that the stress hormones allow the birds to react more quickly to stressors, survive them, and, perhaps, remember them.

The pioneering sparrows also have a taste for new foods. When Martin's graduate student Andrea Liebl tested the birds on foods as foreign to them as freeze-dried strawberries and puppy chow, she found that the sparrows from older, more established populations would have nothing to do with the weird new foods, even when they were very hungry. In contrast, the leading birds—without a moment's hesitation—wolfed down the berries and chow. At the edge of a bird's range, foods and other resources are likely to be novel, explains Liebl. So individuals that are open to trying new things have a big advantage. Otherwise, they might starve.

IF IT'S SO ADVANTAGEOUS to be open to the new and flexible in your eating and foraging habits, why don't all sparrows adopt these traits?

Because they're risky. Flexibility carries costs. Curiosity may kill the bird as well as the cat. Exploring the new and unknown takes time and energy, and it can get you into trouble. Taste a new food and you take the chance of tasting a new toxin or pathogen along with it.

Great blue herons are known for their experimental eating, taking in all kinds of large, cumbersome, or unwieldy prey—snakes, sticklebacks, sculpins, and other spiny fish. But one bird off the coast of Biloxi, Mississippi, lately broke new ground when he branched into the untested realm of elasmobranchs. It was a quiet November day. A group of scientists from the Dauphin Island Sea Lab saw a heron just off the coast striking something beneath the water again and again, without success. Then its head disappeared beneath the surface for a time and it emerged with an Atlantic stingray impaled on its beak. Lots of creatures eat elasmobranchs, including killer whales, fur seals, and a suite of sharks. But birds? The ray "wriggled and whipped its tail and venomous spine back and forth" in the heron's beak, said the scientists. After twelve minutes of struggle, the bird managed to fold the stingray tightly in its mouth, expand its esophagus, and swallow the thing whole, apparently without distress.

A brown pelican found dead on the Baja coast tried the same trick but without success. The bird had the tail spine of a stingray embedded in its throat and had died, presumably, as a result of choking or poisoning. "Proof that opportunism is a hazardous way of life," noted the observers.

Keas, those clever, playful parrots endemic to New Zealand, will consume almost anything. They eat a hundred species of plants, insects, eggs, seabird chicks, and animal carcasses—which may be one reason they survived the mass extinctions wrought by human settlement on the island. They even tasted the sheep introduced into their alpine habitat in the 1860s, at first feeding on dead sheep and then developing a new foraging strategy: riding on the backs of live sheep to feast directly on their fat and muscle tissue.

The same traits that helped keas survive in a harsh environment for most of its evolutionary history have lately put them in harm's way. The sheep-feeding innovation did not endear them to farmers, who placed a bounty on the birds, resulting in the deaths of an estimated 150,000 keas. The birds' investigative tendencies in ski fields, parking lots, and rubbish dumps often endanger the remaining 1,000 to 5,000 individuals. One kea in the alpine village of Mount Cook ran into trouble on account of his skill at opening the lids of trash cans. He was found dead with 20 grams of dark liquid material in his crop. The cause of death? "Methylxanthine toxicity after opportunistic ingestion of dark chocolate."

The point is that it's dangerous to explore the new and unknown. A strategy of seeking out and sampling alternative food or shelter may benefit a house sparrow as it's first settling in, when much of the environment is unfamiliar. But, as Lynn Martin suggests, "Eating new (possibly nasty) stuff increases the risks, including the risk of infection." Once the birds are established in a place, they may shift their strategy and stick with what's known.

Again, it's advantageous to have a mix of personalities: some risk takers to copy (or not, if the behavior is unwise) and some who play it safe.

HERE, THEN, is a recipe for the house sparrow's success:

- A taste for novelty
- A pinch of the innovative
- A dash of daring
- And, perhaps, a penchant for hanging out in mixed gangs

Add to this its love of habitats that have become widespread on this planet and its ability to brood several times in a single breeding season. (The latter, called a bet-hedging strategy, reduces the fitness costs of failed breeding attempts, which, as Daniel Sol says, "appears to be particularly useful in urban environments, where the risk of reproductive failure may be high.") Put it all in the blender, and the upshot is a bird that's a consummate adapter, one that can easily switch to a new food or foraging strategy or an unorthodox nesting site. It's a different breed of genius. In this case, "the measure of intelligence is the ability to change." Darwin didn't say that, but Einstein supposedly did.

THE HOUSE SPARROW is not the only bird that has learned to love garbage and nest in drainpipes. A number of other species—pigeons, crows, a few small songbirds—are synanthropes, well suited to life in radically changed environments such as cities, which are rife with novel opportunities, as well as dangers such as cars, electrical lines, buildings, windows. (In Toronto, for instance, just twenty buildings have been the sites of fatal collisions for more than thirty thousand birds.) Daniel Sol and his colleagues looked at eight hundred species around the world and identified a number that are, as Sol says, "true urban exploiters that have attained higher densities in the cities than in the natural surroundings." These include members of the crow, blackbird, and pigeon families. The team also drummed up a list of the most common traits and behaviors that allow these birds to go to town. Chief among them are big brains and the ability to deal with strange foods, the hazards of traffic, perpetual light, and constant din. Among songbirds, for instance, musical accommodation is key—the will and ability to switch up your tunes. Cities hum, buzz, roar, and rumble in the lower frequencies. Canadian researchers recently found that when traffic noise is high, black-capped chickadees sing their *fee-bee* songs at higher frequencies so that they're heard over the low-frequency urban cacophony. When the din diminishes, they revert to their familiar lower, slower, more musical melody. "The remarkable vocal flexibility exhibited by chickadees may be one reason that they thrive in urban environments," say the researchers. European robins get around urban sound by singing in the quieter night.

Cities have been called learning machines. They may make smart birds even smarter.

WHO CAN'T COPE in the urban jungle? Those birds distinctly unsparrowlike, skittish, or set in their ways. The bird startled off the nest by human hustle and bustle or thrown for a loop by round-the-clock lighting. The small-brained, the inflexible, the specialist.

The same is true for birds that live in farmlands, even those far from cities and suburbs. When scientists looked at trends in bird populations living in farmlands in the United Kingdom over a period of thirty years, they found dramatic declines in the small-brained species such as warblers and tree sparrows, while species with relatively large brains, such as magpies and tits, were still faring well. Birds that were most particular about their habits and their haunts seemed to be suffering the most.

New insights from the farms and jungles of Central America confirm this. For a dozen years, biologists from Stanford University have been counting birds in three different types of habitat in Costa Rica: relatively pristine forest reserves, "mixed" farmlands (with different sorts of crops and dotted with small patches of forest), and, finally, large, intensively farmed single-crop plantations of sugarcane or pineapple.

Over a period of twelve years in forty-four different transects, the team members counted 120,000 birds of five hundred different species. The mixed farmlands, to their surprise, had as many species as the pristine forests. But the scientists were interested in more than simple species diversity. They wanted to know whether there was *evolutionary* diversity—birds from distant branches of the evolutionary tree.

What they found was telling.

In the farmed landscapes, constantly disturbed and worked over by humans, most birds present were members of closely related species that adapt easily to change, primarily sparrows and blackbirds that had evolved as separate species only within the past couple of million years. Absent were representatives from branches far apart from these birds on the evolutionary tree: the great tinamou, for instance, a stocky, speckled, flightless bird that diverged from blackbirds and sparrows some 100 million years ago. This tinamou thrives only in a specialized jungle habitat, where its drab brown and gray plumage blends in with the foliage. (Though its eggs don't necessarily—they're renowned for their high gloss and spectacular colors: lime green, sky blue, and a coppery purple-brown.)

FOR THOSE INTERESTED IN preserving bird diversity, this raises an important question. Do smart, adaptive bird lineages such as sparrows and blackbirds evolve more rapidly, generating more new species? Research by Daniel Sol and his colleagues suggests this may be so. The number of species varies greatly among different groupings of birds. The Passerida (sparrows and related songbirds) has 3,556 species, while the Odontophoridae (quails and their relatives) contains only 6 species. In taxonomic studies, Sol has shown that big-brained species that are innovative, adaptable, and good at invading new environments diversify at a speedier rate. This includes species-rich groups such as the corvids, parrots, and birds of prey, which have a capacity for quickly adjusting their feeding behavior.

It's called the behavioral drive theory. The idea is this: Individual birds that adopt a new habit expose themselves to a new set of selection pressures. These new pressures may favor certain genetic variations or mutations that improve a bird's effectiveness at living in a new way or within a new context. Birds with these variations diverge from the rest of the population. In other words, novel behaviors foster novel traits, which produce new species. Over evolutionary time, then, opportunistic birds that can easily swap one food source for another or use a new foraging technique have generated more species than their less adaptable peers.

This may go a long way toward explaining why there are close to 120 species of corvids and only a handful of ratites, flightless birds like the ostrich and the emu. It also begs the question of whether we humans, in creating novel and unstable environments, are changing the very nature of the bird family tree.

EVEN ON REMOTE MOUNTAIN PEAKS where the forest is pristine, birds of old lineages feel the ripple of human effects—not from the encroachment of cities and farms but from something even more pervasive.

In early 2014, two young researchers, Ben and Alexandra Freeman of Cornell University, found that 70 percent of the bird species living in the mountains of New Guinea—eighty-seven species—had shifted their ranges upslope by an average of five hundred feet or so in the past half century to escape rising temperatures from global warming. Ben Freeman is fascinated that most mountain birds in the tropics live at very narrow bands of elevation. "I find it astonishing that I can walk uphill through forest where a given species is absent, to forest where that species is abundant, and finally to forest that again lacks the species—all in the course of fifteen minutes of hard hiking," he says. This holds true despite the apparent sameness of the forest as he ascends—and the birds' ability to fly to higher or lower elevations. "Is it a Goldilocks scenario, where other elevations are too hot or cold?" he wonders.

It seems so.

On Mount Karimui, an extinct volcano on the main island, the range of the magnificent bird-of-paradise had ascended more than three hundred feet as a result of warming of just 0.7 degree Fahrenheit. "Because a mountain is like a pyramid," says Freeman, "there's less area for habitat available as they move up the mountain. They're being squeezed both by temperatures and for space." The white-winged robin, for instance, which lived on the top one thousand feet of a mountain fifty years ago, now wedges into just the top four hundred feet.

Temperatures in New Guinea are expected to rise another 4.5 degrees Fahrenheit by the end of the century. Four species of birds in search of cooler temperatures have already reached the summit of Mount Karimui and have nowhere else to go. These old-lineage, specialized birds appear to be climbing their way to local extinction. Even a rise of another degree or two will push their thermal zone off that mountain and into the sky.

NOT FAR FROM where I live is a small mountain I like to visit, Buck's Elbow. Nothing exotic like Mount Karimui, just an old Virginia hill. It's a place I go to for perspective and the wide-open views. The summit here is bare, almost like an Irish heath, and on a clear day offers a 360-degree panorama of the Appalachian Mountains around it. But this spring afternoon, the mountaintop is nested in cloud. A blanket of fog has rolled in, enshrouding the summit and muffling sound.

The top of Buck's Elbow was always bald, but the slopes below were once forested in old growth, long ago cut over like so much original forest in the East. I once saw a global map of human impact showing that only around 15 percent of the world's land area has escaped the human footprint. Towns and cities, farms, roads, nighttime light: They're everywhere but that thin slice of earth's pie. And even where they are not—like Mount Karimui—the planet is changing. Estimates put the rise in global temperatures at 3 to 7 degrees Fahrenheit over the next sixty years.

In these parts, everything seems to bloom earlier than it used to. Mayapples open their shy white blossoms in mid-April. Yellow lady's slippers unfurl on the mountain slopes nearly a month before they did in the past.

Just a few days earlier, in a little park not far from here, I spotted a baby eastern bluebird on the limb of a locust tree. Roughly two or three weeks past hatching, he still had that goofy baby bird look— wide gaping mouth, short tail, spiky feathers sticking up from his head. The ornithologist with me was staggered. "It's unheard of to have fledged baby bluebirds around here in April. It's too early for this."

Virginia's climate is moving "down latitude," as they say. According to projections by The Nature Conservancy, the state will be as hot as South Carolina by 2050, as hot as northern Florida fifty years after that. Rising temperatures are changing the schedules of resident birds; they're also shifting the range of temperate species toward the poles. Fifty years ago, "southern" species such as cardinals and Carolina wrens were rare in the northeastern United States; now they're common.

When there is no other place to go, birds cope with rising temperatures in two ways: by evolving or by adjusting their behavior.

Great tits, known for their behavioral flexibility, appear to have sorted this out, at least according to a long-term population study on the tits breeding in Wytham Woods. A team from Oxford showed that the short generation time of the tits allows them to evolve quickly, though not quickly enough. Critical to their survival is their ability to rapidly adjust their behavior. The great tits of those woods time the laying and hatching of their eggs to coincide with the spring peak of the moth caterpillars they feed to their young. The caterpillars emerge from their pupae with the blossoming of trees in spring, the timing of which is dictated by temperature. As temperatures have risen over the past half century, tree blossoming and the caterpillar boom are occurring earlier than they did when the study began in 1960. If the tits were hardwired to lay their eggs at the same time every year, they would miss the caterpillar boom and their young would starve. But the birds have apparently tracked this shift and are now laying their eggs about two weeks earlier.

The scientists' models suggest that this ability to adjust their behavior could allow the birds to survive a warming of 0.9 degree Fahrenheit per year. Without it, the tits would face a five-hundred-fold risk of extinction.

When the researchers used the models to predict how other birds might handle the warming trends, they found that larger, longer-lived species fared worse. These birds have longer generation times, which means that they evolve more slowly, so they're more dependent on changing their behavior to survive. If these projections hold true, it may bode ill for bigger birds that score low on versatility.

Long-distance migrants are particularly susceptible to global warming. These birds are largely small brained and inflexible in their behavior. They depend on precisely timed, once-a-year blooms of food to fuel their breeding. If warming changes the traditional timing of food availability, chances are they'll suffer. Most vulnerable may be birds that breed or winter at high latitudes, where alterations brought about by climate change are expected to be especially severe.

Many migratory birds also depend on precisely timed stopovers for feeding at critical points along their routes. Take the red knot, a bird of modest brain but prodigious travel. Each spring, it journeys ninety-three hundred miles from Tierra del Fuego to the Arctic. For thousands of years the red knot has relied for sustenance on a precisely timed rendezvous with horseshoe crabs laying their eggs on the beaches of the Delaware Bay. The eggs are so packed with fat that a red knot can double its body weight in just ten days of feasting. Since the 1980s, the red knot population has dropped by 75 percent, largely because of overharvesting of horseshoe crabs. The harvesting has slowed of late, but climate change may deal the birds another blow. Crabs and birds must arrive simultaneously if the birds are going to make it to their nesting grounds in the Arctic. Shifting temperatures may throw the red knot out of sync with this food source so crucial to its annual marathon. If warming water temperatures cause the crabs to lay their eggs before the knots arrive, the birds will miss their vital feast.

THE TRUTH IS, relatively intelligent birds are also at risk—the mountain chickadee, for instance, a hardy little bird partial to mountain conifer forest. Its habitat is expected to dwindle by 65 percent over the next half century. Moreover, global warming could in theory change the cognition and brain structure of this chickadee. Recall that chickadees living at higher elevations have bigger brains than do their peers living lower down. According to Vladimir Pravosudov, if the weather is warmer, winter will provide less selection pressure, so the birds may lose their edge, in both hippocampus size and intelligence. "If maintaining better memory has costs," he argues, "'smarter' birds will be at a disadvantage. Also, these populations will be quickly invaded by more southern, not-so-smart birds, which will lead to overall reduction in cognitive ability."

Even the canny and adaptable house sparrow has its limits. In Ben Freeman's home city, Seattle, the 2014 Christmas Bird Count totaled just 225 house sparrows within the city limits. "That's the lowest total ever," says Freeman, "and one piece of evidence that house sparrows may be declining." Indeed, around the globe the bird is experiencing rapid and massive declines—in North America, Australia, and India, but especially in some towns and cities across Europe. Its decline generates few headlines, but the sparrow is now listed as a species of conservation concern in Europe; in Great Britain, it's red listed. In the past half century, the United Kingdom has lost an average of fifty house sparrows every hour. No one is certain why. The survival of nestlings appears to be the problem, perhaps because they're not getting enough food. Gardens converted to parking lots or low insect density because of exotic vegetation or pollution may be contributing, or perhaps it's due to the loss of parent birds to collisions with cars or predation by the growing numbers of domestic cats and city-loving raptors. Some evidence from Israel points the finger at climate change. Lynn Martin says he's skeptical of these theories, but he doesn't have a good explanation to offer in their stead. "I wouldn't rule out disease of some sort," he says. Whatever the cause of their decline, if sparrows are the new canaries, ours is a troubled mine indeed.

I SIT FOR A WHILE in the gray silence. The quiet on Buck's Elbow is so complete I can hear my own breathing. In this gloom, it's hard to fathom the searing power of the sun's rays. But it's possible to imagine something else: songless woods, fields, mountains. The wisdom is that humanity is driving roughly half of all known life to extinction, including one in four species of birds. It appears to be mainly the specialists we're pushing out—the small brained, the particular, the old lineages.

The final paragraph of Ted Anderson's book on house sparrows is this: "As I watch live television news from Baghdad, Gaza, Jerusalem, or Kosovo and hear sparrows chirping in the background, I sometimes wonder what opinion, if any, the house sparrow has about the havoc wreaked by its human hosts."

I wonder, too. In their lifetime, my two daughters may witness birds of all minds ebbing into a sea that exists only in memory.

We don't even know what we're losing. Scientists are still turning up new species: two kinds of hawk-owls in the Philippines in 2012, one thought to be extinct because of widespread deforestation on the island of Cebu; and in 2014, the Sulawesi streaked flycatcher, a diminutive bird with a mottled throat and melodious song, hanging on in patches of tall forest left by farmers; and in 2015, the secretive little Sichuan bush warbler living in the dense brush and tea plantations of central China's mountainous provinces.

Have other unknown species flown the coop before we even found them?

We're still wondering how to define bird intelligence, still leaning heavily on human yardsticks. We can't help but measure other minds by their resemblance to our own. Naturally, we value what we're good at: toolmaking, for instance, over true navigation.

A new study suggests that crows exhibit an ability to grasp analogies— the sort of sophisticated understanding once thought solely the domain of humans and other primates. The experiment involved a pattern-matching game. Researchers trained two hooded crows to choose the card that looked exactly like a sample card, with correct answers rewarded by a mealworm concealed in a cup below the matching card. Then they asked the crows to do something new: to pick a card that didn't match the sample card but had the same pattern. For instance, if the card showed two squares of the same size, the crows had to match a card with two circles of the same size rather than one with, say, two circles of different sizes. The crows spontaneously picked the right card without any training—a premier example of analogical reasoning, say the researchers, one of "our" higher-level forms of thinking.

This is a truly amazing demonstration of humanlike mental powers. But shouldn't we also appreciate the complex cognitive abilities of birds in their own right and not because they look like some aspects of our own? Migratory birds may have small brains, but look at the colossal mental maps they carry. And consider the unique and enduring cultural traditions of songbirds. According to Richard Prum, the origin of song learning and culture in oscine songbirds occurred some 30 million or 40 million years ago, "perhaps even predating the completion of the break up of Gondwanaland," he writes. "Although human culture is possibly 100,000 years old, songbirds have been doing 'aesthetic culture' on a grand scale for tens of millions of years."

We're still sorting out why some bird species seem smarter than others. Because they've had to solve problems around them, ecological, technical, or social? Because they've had to sing their hearts out or whip up a beautiful bower to win over a choosy mate?

Intelligence as we understand it may vary among birds, but no bird is truly stupid. As the ornithologist Richard F. Johnston said, "Everything that is is adaptive." Not miraculously, not flawlessly, but with its own kind of genius. And that includes the great tinamou and the kagu. I think back to my encounter with that kagu in New Caledonia, heart thumping, camera dangling from my wrist. The ghostly bird, I've since learned, has large laser-red eyes that help it spot prey in dim forest light. It raises just one chick a year. Once dogs were introduced, this reproductive pattern almost doomed it. But is the kagu really that much dafter than the mockingbird that perched on Jefferson's shoulder to take food from his lips? A species poorly adapted to a new predator is not necessarily stupid. What we see as stupidity in the kagu may be something more like ecological naïveté, reflecting the bird's long-term adaptation to a once benign island environment. "If you evolved with no predators and your food is found around ground level through direct pecking, your cognition will focus on food detection and accurate pecking rather than opportunistic foraging," explains Gavin Hunt. "Who knows why kagus often approach people and dogs? Perhaps it's because they don't like other kagus on their patch of ground, so they investigate new arrivals and potential competitors." But now there are predators around. The kagu's world has changed, and the unavoidable truth for this bird and other old-timers is that their luck may be running out.

It would be easy to give up on these birds, write them off as the collateral damage of human "progress." But as one of the scientists who studied those Costa Rican farms and jungles said, "Having just sparrowlike birds in an ecosystem is like investing only in technology stocks." When the bubble bursts, you lose.

IN THE GLOOM on Buck's Elbow, there's a new, diffuse sort of light that makes the haze seem somehow illuminated from within. Suddenly, I hear a strange whooshing noise from nearby. Three wild turkeys burst out of the fog and dash across the meadow before me, striding long-legged through the tall grasses like the small dinosaurs they are, then vanishing again like magic into the mist. A new study comparing the genomes of birds suggests that, genetically speaking, the turkey is closer to its dinosaur ancestors than any other bird is; its chromosomes have undergone fewer changes than other birds since the days of feathered dinosaurs. Watching the gobblers steal away through the long grass, this is easy to believe.

We almost lost our wild turkeys to the dinner platter in the past century. Arthur Cleveland Bent, writing in the 1930s, claimed that the few survivors had developed a high degree of shrewdness and cunning and gave an example noted by one Dr. J. M. Wheaton in 1882: "As if aware that their safety depended on their preserving an incognito when observed, they effect the unconcern of their tame relatives so long as a threatened danger is passive or unavoidable. I have known them to remain quietly perched upon a fence while a team passed by; and one occasion knew a couple of hunters to be so confused by the actions of a flock of five, which deliberately walked in front of them, mounted a fence, and disappeared leisurely over a low hill before they were able to decide them to be wild. No sooner were they out of sight, than they took to their legs and then to their wings, soon placing a wide valley between them and their now amazed and mortified pursuers."

Not all the news is bad. The wild turkey's numbers have since recovered, and now they're cropping up in greater numbers in every state except Alaska. Here they are enthusiasts of the oak and beech forests cloaking the mountain slopes. Like kagus, they are ground foragers. And like kagus, they are not considered the sharpest tacks in the drawer—Dr. Wheaton's story notwithstanding. But even a bird apparently short on brains can be big on presence. As Aldo Leopold reminds us in his riff on the physics of beauty, "The autumn landscape in the north woods is the land, plus a red maple, plus a ruffed grouse. In terms of conventional physics, the grouse represents only a millionth of either the mass or the energy of an acre. Yet subtract the grouse and the whole thing is dead."

The planet has had catastrophic species loss in the past. Out of cataclysmic events may come new creatures. Evidence suggests that the "big bang" radiation of species that occurred for songbirds, parrots, pigeons, and other birds took place after the mass extinction event on earth about 66 million years ago that killed off the dinosaurs. On the scale of deep time, the "sixth mass extinction" may be one of many such events. But the measure that matters most to most of us is the measure of a human lifetime. It isn't necessarily comforting to think that nature may bounce back in a few million years or so. Moreover, while there may be even more than ten thousand species of birds down the evolutionary road, they won't be randomly descended from the species present today. Half may be from the genus *Corvus*, suggests Louis Lefebvre. "People may be unhappy with that idea," Lefebvre told me. "They think crows are plain and ornery. But who knows? In two million years, they could be colorful, beautiful singers."

True. But who will be there to hear them sing? In the meantime, will we settle for diminishment, diversity reduced to sparrowlike species that play by our rules? Or will we strive to preserve the broadest possible swath of the avian tree of life, big brained and small brained, specialist and generalist, old species and new?

"AS A HUMAN BEING," Einstein once wrote, "one has been endowed with just enough intelligence to be able to see clearly how utterly inadequate that intelligence is when confronted with what exists."

We're still teasing out whether it pays for a bird to be smart—how and why and under what conditions intelligence may increase fitness. Do brighter individuals reproduce more? Oddly enough, evidence is scarcer than hen's teeth. "Measuring the fitness benefits of a given trait in the real world is never easy, regardless of the trait," writes Sue Healy. Figuring out the link between bird cognition and fitness is a kind of golden goose in the field. It's tricky because the fitness benefits of a trait such as behavioral flexibility may be evident only in particular situations, says Daniel Sol—years when food is scarce, for instance. Under favorable conditions, the specialist may do better. (It's not unlike the findings about the finches of the Galápagos: In some years, large beaks are adaptive; in other years, smaller beaks fare better.)

There are trade-offs. Daniel Sol has data suggesting the existence of a trade-off between fecundity and survival. In general, smaller-brained birds (which tend to be short-lived) have a larger clutch size, while larger-brained birds (which are usually longer-lived) have a smaller clutch size. But the bigger-brained birds often have a higher rate of survival. It's a balancing act. "Larger-brained birds have a slow life strategy, where the energy goes to survival rather than to reproduction," Sol explains. "A long reproductive life can increase the productivity of these slow-living species—but they will never achieve the high productivity of fast-living species that prioritize reproduction over survival." On the other hand, he says, "a fast-life strategy can favor rapid population growth when conditions are favorable but can be risky if they are not. When there are good and bad years, it may pay to have a slow-life strategy, particularly if the bird has cognitive adaptations to survive in the bad years." In other words, says Sol, "both strategies, living fast or living slow, can be more or less useful depending on the environment."

What about within a species? Do more ingenious individuals have more young? There's conflicting evidence. One study of wild great tits on the island of Gotland in Sweden showed that parents that were speedier at a problem-solving task (pulling a string to open a trapdoor to their nesting box) had a higher nestling survival rate than parents who couldn't solve the task. They had bigger clutches, hatched more eggs, and fledged more nestlings.

However, in a close look at breeding pairs of great tits in Wytham Woods, Ella Cole of Oxford and her colleagues found that things may not be so simple. "Smarter" birds—those that quickly solved a puzzle that required pulling a stick out of a feeder to release tasty treats—laid more eggs and foraged more efficiently, but they were also more likely to abandon their nests. In the end, it was a reproductive wash. In the wild, natural selection does not seem to favor problem-solver great tits over nonsolvers, say the Oxford researchers. The problem solvers may produce bigger clutches because they're better at exploiting their environment, but they also tend to be more skittish about predators and more quickly abandon a nest. (The same has been found in mountain chickadees. The brainier ones that live higher up than their fellow lower-elevation chickadees desert their nests more frequently.)

There could be a catch, though. As the scientists suggest in their paper, the nest failures for the tits may have been due to the experimenters trying to ring the chicks at too young an age. "Were the good problem solvers, then, simply more sensitive to this artificial-experimenter disturbance compared with the nonsolvers, and deserted their nests more often because of it?" asks Neeltje Boogert. "It would be very interesting to test whether these good problem solvers are also more sensitive to real predators, and thus more likely to abandon their nests, as the authors suggest," says Boogert. Without this experimenter-disturbance factor, would the study have confirmed a positive relationship between problem-solving performance and reproductive success? The uncertainty goes to show how challenging it is to conduct these sorts of studies and how difficult it is to tease out all the variables.

IN ANY CASE, we may think of intelligence as universally advantageous, but it isn't always. There are trade-offs with any trait, including being a quick study and a fast learner. Bold birds that respond quickly to problems may suffer an accuracy trade-off as a result of their speed. On the island of Barbados, for instance, Simon Ducatez found that some Carib grackles are speedy problem solvers, while others are slower. In the end, the quicker wits tend to do worse on tests such as reversal learning (just as the Bajan bullfinches did) than the slower, more accurate problem solvers. "Bolder individuals tend to explore more quickly but more superficially," explains Daniel Sol; "the slower explorers gain better information and use this to act with more flexibility." Why do both types persist within a population? "Perhaps different types do better in their environment in different years," speculates Ducatez, which may explain why cognitive ability varies from bird to bird. And why, as the sparrows have taught us, it pays to have a mix.

THE FOG IS LIFTING. I can begin to make out the undulant curtain of the Blue Ridge Mountains across the valley, purpled by the haze. From a grove of trees nearby comes the piercing *zeet* of a chickadee. I wander over, and there is the bird perched in a pine tree, rolling out its string of *dee*s, perhaps taking measure of my presence. One has only to consider the extraordinary genius packed tightly into that tiny puff of feathers to lay the mind wide open to the mysteries of a bird's knowing—the what and the why. These are wonderful puzzles to keep around on our intellectual bookshelf, to remind us how little we still know.

ACKNOWLEDGMENTS

I can't begin to thank adequately those who helped me make this book.

I relied on the research of numerous scientists who have dedicated their lives to studying birds and their brains. Their names, and my debts to them, fill the book.

I'm especially grateful to the following ornithologists, biologists, psychologists, and animal behavior specialists, who gave generously of their time and knowledge while I was researching the project. Louis Lefebvre of McGill University opened his lab to me at the Bellairs Research Institute in Barbados and over the course of several days led me through the world of bird cognition, explaining his own research, offering thoughtful insights into the field as a whole, and answering my innumerable questions with patience, eloquence, and humor. He also read an early draft of the complete manuscript and provided helpful comments and suggestions. During my stay at Bellairs, Lima Kayello, Jean-Nicolas Audet, and Simon Ducatez were uniformly generous in sharing their research and ideas.

When I visited New Caledonia, Alex Taylor of the University of Auckland spent many hours kindly and thoughtfully showing me various aspects of his work with the crows and offering his expertise on bird cognition. Elsa Loissel provided informative conversation, contacts, and companionship while hiking with me through the Parc des Grandes Fougères; she also supplied superb photographs of the kagus we saw together, as well as landscapes of New Caledonia and its crows.

Many other busy but generous individuals spent time talking with me, supplying me with references, and reading and rereading drafts of sections of the book exploring their work, including Lucy Aplin of Oxford University; Gerald Borgia of the University of Maryland; John Endler of Deakin University in Victoria, Australia; Stephen Brusatte of Edinburgh University; Jon Hagstrum of the U.S. Geological Survey; Richard Holland of Queens University, Belfast; Gavin Hunt of the University of Auckland; Erich Jarvis of Duke University; Jason Keagy of Michigan State University; Vladimir Pravosudov of the University of Nevada; Amanda Ridley of the University of Western Australia; and Daniel Sol of Spain's Centre for Research on Ecology and Forestry Applications.

Russell Gray of the University of Auckland kindly shared with me the videos of his brilliant Nijmegen lectures at the Max Planck Institute for Psycholinguistics in 2014.

I owe a large debt of gratitude to Neeltje Boogert of the University of St Andrews, who generously brought her scientific and editorial eye to bear on much of the manuscript, reading with great care and intelligence, some parts more than once. She made better every page she handled.

Many other scientists from all over the world read portions of the manuscript and offered corrections on matters of scientific fact, rescuing me from all kinds of inky embarrassment. Warm thanks on this account to the following:

In the United States: Arkhat Abzhanov of Harvard University; Carlos Botero of Washington University; Nancy Burley of the University of California, Irvine; Lainy Day of the University of Mississippi; Judy Diamond of the University of Nebraska; Ben Freeman of Cornell University; Luke Frishkoff of Stanford University; Tim Gentner of the University of California, San Diego; Walter Herbranson of Whitman College; Lucia Jacobs of the University of California, Berkeley; Alan Kamil of the University of Nebraska; Marcy Kingsbury of Indiana University; Sarah London of the

University of Chicago; Lynn ("Marty") Martin of the University of South Florida, Tampa; John Marzluff of the University of Washington; Shigeru Miyagawa of the Massachusetts Institute of Technology; Richard Mooney of Duke University; Gail Patricelli of the University of California, Davis; Irene Pepperberg of Harvard University; Lauren Riters of the University of Wisconsin; and Rhiannon J. D. West of the University of New Mexico.

In the United Kingdom: Nicola Clayton of Cambridge University; Sue Healy of the University of St Andrews; Richard Holland of Queens University, Belfast; Laura Kelley of Cambridge University; Ljerka Ostojić of Cambridge University; Christian Rutz of the University of St Andrews; Murray Shanahan of Imperial College London; and Chris Templeton of the University of St Andrews.

In Europe: Alice Auersperg of the University of Vienna; Johan Bolhuis of Utrecht University; Jenny Holzhaider of Gräfelfing, Germany; Henrik Mouritsen of the University of Oldenburg; Andreas Nieder of the University of Tübingen; Niels Rattenborg of the Max Planck Institute for Ornithology; and Sabine Tebbich of the University of Vienna.

In Australia and New Zealand: Russell Gray, Gavin Hunt, and Alex Taylor of the University of Auckland; and Teresa Iglesias of Macquarie University in Australia.

And elsewhere: Laure Cauchard of the University of Montréal; Suzana Herculano-Houzel of the Federal University of Rio de Janeiro, Brazil; Kazuo Okanoya of the University of Tokyo; and Shigeru Watanabe of Keio University in Tokyo.

The comments and criticisms of these experts were hugely important in getting me back on track where I had wobbled off the rails. Any lingering mistakes that may still lurk in these pages are clearly and completely mine.

Many friends and colleagues offered invaluable help or buoyed my spirits with their interest in my work. When I overheard Karin Bendel telling a friend about her African grey parrot, Throckmorton, she met my curiosity with kindness and generously shared stories of Throckmorton and of her pet cockatiel, Isabeau. Barrie Pollock did the same, offering tales of her grey, Alfie. Michele and Joey Mangham let me spend an afternoon on their wool farm in the company of Luke, Joey's monk parakeet, who obligingly sat on my shoulder and periodically leaned in to my ear to say, "Whisper, whisper, whisper."

Daniel Bieker, a gifted teacher and ornithologist, guided my fellow field ornithology students and me on outings (many of which found their way into this book) and gave me a refined appreciation for birdsong. He also read the entire manuscript with an eye to the accuracy of the bird observations. David White, an experienced birder, shared stories, humor, and expertise.

My dear friend Miriam Nelson has had a hand in many of my books, sometimes as a colleague or coauthor but more often for no other reason than generosity and friendship. She read an early draft of the manuscript for this book and made many excellent suggestions. Several friends offered encouragement and ideas (and sometimes little bird videos), especially Susan Bacik, Ros Casey, Sandra Cushman, Laura Delano (who shared her "peacock in a mistral" story), Liz Denton, Mark Edmundson, Dorrit Green, Sharon Hogan, Donna Lucey, Debra Nystrom, Dan O'Neill, Michael Rodemeyer, John Rowlett, Nancy Murphy-Spicer, David Eddy Spicer, Henry Wiencek, and Andrew Wyndham. Heartfelt thanks to them and also to my loving and generous father and stepmother, Bill and Gail Gorham, and to my beloved sisters, Sarah Gorham, Nancy Haiman, and Kim Umbarger, for their warm support and interest in my work. A shout-out as well to my two dear and thoughtful daughters, Zoë and Nell, for their constant love and encouragement, and for keeping me—and my office—in birds. ("Put a bird on it!")

For more than two decades, I've had the great honor and joy of working with my agent, Melanie Jackson. It's hard to imagine writing a book without the benefit of her enthusiasm, intelligence, and good judgment in all matters. I am also extremely fortunate to have as my editor Ann Godoff, and am deeply grateful for her gifted editorial vision and her abundant help with this book. Thanks, too, to Sofia Groopman and Casey Rasch for their assistance in shepherding the book through the publication process, and to John Burgoyne for his lovely illustrations and the pleasure of collaborating.

Finally, my profound love and gratitude to my dear Karl, who in fact comes first in every way because he has flown by my side all these years through the gusts and squalls of life and work; without his encouragement, wisdom, patience, support, companionship, perspective, humor, and love, none of this would be.

NOTES

Introduction: The Genius of Birds

2 **Beginning in the 1980s:** The following information about Alex comes from I. M. Pepperberg, *The Alex Studies* (Cambridge, MA: Harvard University Press, 1999); I. M. Pepperberg, "Evidence for numerical competence in an African grey parrot (*Psittacus erithacus*)," *J Comp Psych* 108 (1994): 36–44; I. M. Pepperberg, "Ordinality and inferential abilities of a grey parrot (*Psittacus erithacus*)," *J Comp Psych* 120, no. 3 (2006): 205–16; I. M. Pepperberg and S. Carey, "Grey parrot number acquisition: The inference of cardinal value from ordinal position on the numeral list," *Cognition* 125 (2012): 219–32.

2 **Until Alex, we thought:** The chimpanzee Washoe understood many words but could not speak—though she learned about 130 signs.

3 **In the 1990s:** G. R. Hunt, "Manufacture and use of hook-tools by New Caledonian crows," *Nature* 379 (1996): 249–51; G. R. Hunt and R. D. Gray, "Species-wide manufacture of stick-type tools by New Caledonian crows," *Emu* 102 (2002): 349–53; G. R. Hunt and R. D. Gray, "Diversification and cumulative evolution in tool manufacture by New Caledonian crows," *Proc R Soc B* 270 (2003): 867–74.

3 **"Can you get the food":** A. A. S. Weir et al., "Shaping of hooks in New Caledonian crows," *Science* 297, no. 5583 (2002): 981.

3 **and some birds brains:** S. Olkowicz et al., "Complex brains for complex cognition—neuronal scaling rules for bird brains" (poster presentation at the Society for Neuroscience annual meeting in Washington, D.C., November 15–19, 2014); Suzana Herculano-Houzel, personal communication, January 14, 2015.

4 **Like our brains:** L. Rogers, "Lateralisation in the avian brain," *Bird Behav* 2 (1980): 1–12.

4 **Magpies can recognize:** H. Prior et al., "Mirror-induced behavior in the magpie (*Pica pica*): Evidence of self-recognition," *PLoS Biol* 6, no. 8 (2008): e202, doi:10.1371/journal.pbio.0060202.

4 **Western scrub jays use Machiavellian:** U. Grodzinski et al., "Peep to pilfer: What scrub-jays like to watch when observing others," *Anim Behav* 83 (2012): 1253–60.

4 **These birds seem to have:** N. S. Clayton et al., "Social cognition by food-caching corvids: The western scrub-jay as a natural psychologist," *Phil Trans Roy Soc B: Biol Sci* 362, no. 1480 (2007): 507–22.

4 **They can also remember:** N. S. Clayton and A. Dickinson, "Episodic-like memory during cache recovery by scrub jays," *Nature* 395 (1998): 272–74; N. S. Clayton et al., "Episodic memory," *Curr Biol* 17, no. 6 (2007): 189–91.

4 **This ability to remember:** L. Cheke and N. S. Clayton, "Mental time travel in animals," *Wiley Interdiscip Rev Cogn Sci* 1, no. 6 (2010): 915–30.

4 **News has arrived that:** R. O. Prum, "Coevolutionary aesthetics in human and biotic artworlds," *Biol Phil* 28, no. 5 (2013): 811–32.

4 **In 2015 researchers:** R. Rugani et al., "Number-space mapping in the newborn chick resembles humans' mental number line," *Science* 347, no. 6221 (2015): 534–36.

5 **Baby birds:** R. Rugani et al., "The use of proportion by young domestic chicks," *Anim Cogn* 13, no. 3 (2015): 605–16; R. Rugani et al., "Is it only humans that count from left to right?," *Biol Lett* (2010), doi:10.1098/rsbl.2009.0960.

5 **They can do simple arithmetic:** R. Rugani, "Arithmetic in newborn chicks," *Proc R Soc B* (2009), doi:10.1098/rspb.2009.0044.

5 **"A man would be worn out":** L. Halle, *Spring in Washington* (Baltimore: Johns Hopkins University Press, 1988), 182.

7 **Like the black-billed cuckoo:** Observation of ornithologist Dan Bieker.

8 **"the capacity to learn":** W. F. Dearborn, quoted in R. J. Sternberg, *Handbook of Intelligence* (Cambridge: Cambridge University Press, 2000), 8.

8 **"the capacity to acquire capacity":** H. Woodrow, quoted in R. J. Sternberg, *Handbook of Intelligence* (Cambridge: Cambridge University Press, 2000), 8.

8 **"Intelligence is what is measured":** E. G. Boring, "Intelligence as the tests test it," *New Republic* 35 (1923): 35–37.

8 **"There seem to be almost":** R. J. Sternberg, "People's conceptions of intelligence," *J Pers Soc Psych* 41, no. 1 (1981): 37–55.

8 **As a class, birds:** I'm referring here to *Aves*, the "crown group" of birds, i.e., the living species and all descendants of their most recent common ancestor. Feathered winged animals that fly have been around for more than 150 million years. E. D. Jarvis et al., "Whole-genome analyses resolve early branches in the tree of life of modern birds," *Science* 346, no. 6215 (2014): 1320–31; S. Brusatte et al., "Gradual assembly of avian

body plan culminated in rapid rates of evolution across the dinosaur-bird transition," *Curr Biol* 24, no. 20 (2014): 2386–92.

8   **In the late 1990s:** K. J. Gaston and T. M. Blackburn, "How many birds are there?" *Biodivers Conserv* 6, no. 4 (1997): 615–25.

9   **insight, for instance:** Thorpe defined *insight* as "the sudden production of a new adaptive response not arrived at by trial behaviour or as the solution of a problem by the sudden adaptive reorganization of experience." W. H. Thorpe, *Learning and Instinct in Animals* (London: Methuen & Co. Ltd., 1964), 110.

10 **Some scientists call these building blocks:** A. Taylor, "Corvid cognition," *WIREs Cogn Sci* (2014), doi: 10.1002/wcs.1286; Alex Taylor, personal communication, May 2014; R. Gray, "The evolution of cognition without miracles" (Nijmegen Lectures, January 27–29, 2014), video recording at http://www.mpi.nl/events/nijmegen-lectures-2014/lecture-videos.

10 **More recently, *genius*:** Defined in 1901 by British novelist Amelia Barr in her essay, "A successful novelist: Fame after fifty," in O. Swett Marden, *How They Succeeded: Life Stories of Successful Men Told by Themselves* (Boston: Lothrop Publishing Company, 1901), 311.

11 **The classic example:** J. B. Fisher and R. A. Hinde, "The opening of milk bottles by birds," *Br Birds* 42 (1949): 347–57; L. M. Aplin et al., "Milk-bottles revisited: Social learning and individual variation in the blue tit (*Cyanistes caeruleus*), "*Anim Behav* 85 (2013): 1225–32.

12 **To meet the challenges of filter feeding:** John Endler, personal communication, February 3, 2015.

12 **"Again and again":** Ibid.

12 **"the nearest analogy to language":** C. Darwin, *The Descent of Man* (London: John Murray, 1871), 59.

12 **A group of two hundred scientists:** A. R. Pfenning et al., "Convergent transcriptional specializations in the brains of humans and song-learning birds," *Science* 346, no. 6215 (2014): 1256846.

14 **A recent report from Audubon tells:** http://climate.audubon.org/article/audubon-report-glance.

One: From Dodo to Crow: *Taking the Measure of a Bird Mind*

17 **And more recently by a wizard:** https://www.youtube.com/watch?v=AVaITA7eB ZE#t=51.

17 **The puzzle was created by Alex Taylor:** The puzzle was an expansion of a three-step experiment in metatool use. A. H. Taylor et al., "Spontaneous metatool use by New Caledonian crows," *Curr Biol* 17, no. 17 (2007): 1504–7.

18 **Spontaneously aiming a tool:** Ibid.

18 **"It suggests that the crows":** Alex Taylor, personal communication, January 7, 2015.

20 **To explore these questions:** L. Lefebvre, "Feeding innovations and forebrain size in birds" (AAAS presentation, February 21, 2005, part of the symposium "Mind, Brain and Behavior"). All quotes and information from Louis Lefebvre are from interviews in Holetown, Barbados, February 26–March 1, 2012.

21 **"depauperate avifauna":** P. A. Buckley et al., *The Birds of Barbados*, British Ornithologists' Union, Checklist Number 24 (2009), 58.

21 **This is partly due to its physical nature:** P. A. Buckley and F. G. Buckley, "Rapid speciation by a Lesser Antillean endemic, Barbados bullfinch, *Loxigilla barbadensis*," *Bull BOC* 124, no. 2 (2004): 108–23.

22 **In Carib grackles:** J. Morand-Ferron et al., "Dunking behavior in Carib grackles, *Anim Behav* 68 (2004): 1267–74.

22 **"There's a cost/benefit ratio":** J. Morand-Ferron and L. Lefebvre, "Flexible expression of a food-processing behavior: Determinants of dunking rates in wild Carib grackles of Barbados," *Behav Process* 76 (2007): 218–21.

23 **However, in** *The Descent of Man*: C. Darwin, *The Descent of Man*.

23 **For Darwin:** C. Darwin, *The Formation of Vegetable Mould Through the Action of Worms* (London: John Murray, 1883), 93.

23 **"Those who are in anthropodenial":** F. B. M. de Waal, "Are we in anthropodenial?" *Discover* 18, no. 7 (1997): 50–53. As de Waal points out, concern about the perils of anthropomorphism appears to be less of an issue in non-Western cultures, where the distinction between humans and nonhumans is not so categorical. See F. B. M. de Waal, "Silent invasion: Imanishi's primatology and cultural bias in science," *Anim Cogn* 6 (2003): 293–99.

23 **Animal cognition is generally:** S. J. Shettleworth, *Cognition, Evolution, and Behavior*, 2nd ed. (New York: Oxford University Press, 2010), 23.

24 **In this view:** R. Samuels, "Massively modular minds: Evolutionary psychology and cognitive architecture," in *Evolution and the Human Mind: Modularity, Language and Meta-Cognition*, ed. P. Carruthers and A. Chamberlain (Cambridge: Cambridge University Press, 2000), 13–46; S. J. Shettleworth, *Cognition, Evolution and Behavior*, 23.

24 **Lefebvre, on the other hand:** S. M. Reader et al., "The evolution of primate general and cultural intelligence," *Philos Trans R Soc Lond B* 366 (2011): 1017–27; L. Lefebvre, "Brains, innovations, tools and cultural transmission in birds, non-human primates, and fossil hominins," *Front Hum Neurosci* 7 (2013): 245.

24 **In his theory:** H. Gardner, "Reflections on multiple intelligences: Myths and messages," *Phi Delta Kappan* 77, no. 3 (1995): 200–209.

25 **A group of fifty-two researchers who assembled:** L. S. Gottfredson, "Mainstream science on intelligence: An editorial with 52 signatories, history, and bibliography," *Intelligence* 24, no. 1 (1997): 13–23; see also I. J. Deary et al., "The neuroscience of human intelligence differences," *Nat Rev Neuro* 11 (2010): 201–11.

27 **"One explanation for this evolutionary shift":** It might also have something to do with the fact that Barbados bullfinch males contribute more parental care than their more colored cousins on the other islands. According to a new paper by Lefebvre and his colleagues, "In birds, species tend to be monochromatic when both sexes participate in parental duties, including nest building. . . . Male Barbados bullfinches, compared to male Lesser Antillean bullfinches, contribute more to nest building, stay longer in the vicinity of their nest after construction and throughout brood rearing, feed females more often and are more aggressive around their nest. . . . Breeding system might thus be an important factor in the loss of male dimorphism in this species." See J. L. Audet et al., "Morphological and molecular sexing of the monochromatic Barbados bullfinch, *Loxigilla barbadensis*," *Zool Sci* 10, no. 31 (2014): 687–91.

28 **Of the thirty bullfinches:** L. Kayello, "Opportunism and cognition in birds" (master's thesis, McGill University, 2013), 55–67.

29 **In a comparable study:** S. E. Overington et al., "Innovative foraging behaviour in birds: What characterizes an innovator?" *Behav Process* 87 (2011): 274–85.

29 **"They circle; now dense":** E. Selous, *Bird Life Glimpses* (London: G. Allen, 1905), 141.

29 **"They must think collectively":** E. Selous, *Thought-Transference (or What?) in Birds* (New York: Richard R. Smith, 1931).

30 **We've since learned that:** I. D. Couzin and J. Krause, "Self-organization and collective behavior in vertebrates," *Adv Stud Behav* 32 (2003): 1–75; I. Couzin, "Collective minds," *Nature* 445 (2007): 715; C. K. Hemelrijk et al., "What underlies waves of agitation in starling flocks," *Behav Ecol Sociobiol* (2015), doi: 10.1007/s00265-015-1891-3.

30 **Instead, each bird is interacting:** I. Lebar Bajec and F. H. Heppner, "Organized flight in birds," *Anim Behav* 78, no. 4 (2009): 777–89; M. Ballerini et al., "Interaction ruling animal collective behavior depends on topological rather than metric distance: Evidence from a field study," *PNAS* 105, no. 4 (2008): 1232–37; A. Attanasi et al., "Information transfer and behavioural inertia in starling flocks," *Nat Phys* 10 (2014): 691–96.

31 **"Unfortunately it is extremely difficult":** Neeltje Boogert, personal communication, April 3, 2015.

32 **It's a notion:** H. Kummer and J. Goodall, "Conditions of innovative behaviour in primates," *Philos Trans R Soc Lond B* 308 (1985): 203–14.

33 **Lefebvre published his observation:** L. Lefebvre and D. Spahn, "Gray kingbird predation on small fish (*Poecillia spp*) crossing a sandbar," *Wilson Bull* 99 (1987): 291–92.

34 **They were seen chipping holes:** T. G. Grubb and R. G. Lopez, "Ice fishing by wintering bald eagles in Arizona," *Wilson Bull* (1997): 546–48.

34 **Once the anecdotes were gathered:** L. Lefebvre et al., "Feeding innovations and forebrain size in birds," *Anim Behav* 53 (1997): 549–60.

34 **What are the smartest:** L. Lefebvre, "Feeding innovations and forebrain size in birds" (AAAS presentation, February 21, 2005, part of the symposium "Mind, Brain and Behavior").

34 **In most cases, there was a correlation:** L. Lefebvre et al., "Feeding innovations and forebrain size in birds," *Anim Behav* 53 (1997): 549–60; S. Timmermans et al., "Relative size of the hyperstriatum ventrale is the best predictor of innovation rate in birds," *Brain Behav Evol* 56 (2000): 196–203.

35 **"After all, the little stint":** Louis Lefebvre, personal communication, January 13, 2015.

35 **Honeybees, with a brain:** R. Menzel et al., "Honey bees navigate according to a maplike spatial memory," *PNAS* 102, no. 8 (2005): 3040–45; M. Marine Battesti et al., "Spread of social information and dynamics of social transmission within drosophila groups," *Curr Biol* 22 (2012), 309–13, doi:10.1016/j.cub.2011.12.050.

36 **The ratio of brain size:** See D. M. Alba, "Cognitive inferences in fossil apes (Primates, Hominoidea): Does encephalization reflect intelligence?," *J Anthropol Soc* 88 (2010): 11-48; R. O. Deaner et al., "Overall brain size, and not encephalization quotient, best predicts ability across non-human primates," *Brain Behav Evol* 70 (2007): 115–24.

36 **"Look, if you want to understand":** E. Kandel's anecdote quoted in C. Dreifus, "A Quest to Understand How Memory Works: A Conversation with Eric Kandel," *New York Times*, Science Times, March 6, 2012.

Two: The Bird Way: *The Avian Brain Revisited*

39 **"I have seen a chickadee":** E. H. Forbush, *Useful Birds and Their Protection* (Aurora, CO: Bibliographical Research Center, 2010; originally published in 1913), 195.

40 **"a bird masterpiece":** E. H. Forbush, *Natural History of the Birds of Eastern and Central North America* (Boston: Houghton Mifflin, 1955), 347.

40 **Lately, the high:** T. M. Freeberg and J. R. Lucas, "Receivers respond differently to chick-a-dee calls varying in note composition in Carolina chickadees, *Poecile carolinensis*," *Anim Behav* 63 (2002): 837–45.

40 **Chris Templeton and his colleagues:** C. Templeton et al., "Allometry of alarm calls: Black-capped chickadees encode information about predator size," *Science* 308 (2005): 1934–37.

40 **"deep-rooted self-confidence":** In the words of Edward Forbush: E. H. Forbush, *Natural History of the Birds of Eastern and Central North America*, 347.

41 **Chris Templeton once watched:** Chris Templeton, personal communication, February 12, 2015.

41 **Templeton discovered that:** C. N. Templeton, "Black-capped chickadees select spotted knapweed seedheads with high densities of gall fly larvae," *Condor* 113, no. 2 (2011): 395–99.

41 **They stash seeds and other food:** T. C. Roth et al., "Evidence for long-term spatial memory in a parid," *Anim Cogn* 15, no. 2 (2011): 149–54.

42   **A chickadee weighs:** L. S. Phillmore et al., "Annual cycle of the black-capped chickadee: Seasonality of singing rates and vocal-control brain regions," *J Neurobiol* 66, no. 9 (2006): 1002–10.

42   **Bird brains range in size:** A. N. Iwaniuk and J. E. Nelson, "Can endocranial volume be used as an estimate of brain size in birds?" *Can J Zool* 80 (2002): 16–23.

42   **That's the same size as the brain:** N. E. Emery and N. S. Clayton, "The mentality of crows: Convergent evolution of intelligence in corvids and apes," *Science* 306, no. 5703 (2004): 1903–7.

42   **It has double:** Louis Lefebvre, personal communication, January 13, 2015.

43   **so the chickadee could beat:** C. H. Greenewalt, "The flight of the black-capped chickadee and the white-breasted nuthatch," *Auk* 72, no. 1 (1955): 1–5.

43   **Neurons may be small:** S. B. Laughlin et al., "The metabolic cost of neural information," *Nat Neurosci* 1, no. 1 (1998): 36–41.

43   **"Ironically the power of flight":** P. Matthiessen, *The Wind Birds* (New York: Viking, 1973), 45.

43   **For a small bird such as a finch:** R. L. Nudds and D. M. Bryant, "The energetic cost of short flights in birds," *J Exp Biol* 203 (2000): 1561–72.

43   **(By comparison, swimming for a waterbird):** P. J. Butler, "Energetic costs of surface swimming and diving of birds," *Physiol Biochem Zool* 73, no. 6 (2000): 699–705.

43   **To meet the constraints of flight:** General information on bird anatomy and physiology comes from F. B. Gill, *Ornithology* (New York: Freeman, 2007), 141–73.

43   **A bird's bones are dense:** E. R. Dumon, "Bone density and the lightweight skeletons of birds," *Proc R Soc B* 277 (2010): 2193–98.

43   **(So powerful is the downstroke):** D. Lentink et al., "In vivo recording of aerodynamic force with an aerodynamic force platform: From drones to birds," *J Roy Soc Interface* (2015), doi: 10.1098/rsif.2014.1283.

43   **When biologists examined:** G. Zhang et al., "Comparative genomics reveals insights into avian genome evolution and adaptation," *Science* 346, no. 6215 (2014), 1311–19.

43   **The paradoxical result sometimes:** R. C. Murphy, *Oceanic Birds of South America* (New York: Macmillan, 1936).

43   **Evolution has found other:** P. R. Ehrlich et al., "Adaptations for Flight," 1988, https://web.stanford.edu/group/stanfordbirds/text/essays/Adaptations.html; F. B. Gill, *Ornithology* (New York: Freeman, 2007), 115–37.

44   **A bird's wild knot:** J. C. Welty, *The Life of Birds* (Philadelphia: Saunders, 1975), 112.

44   **Its "flow-through" lung:** H. R. Duncker, "The lung air sac system of birds," *Adv Anat Emb Cell Biol* 45 (1971): 1–171.

44   **Birds have the smallest genomes:** E. D. Jarvis et al., "Whole-genome analyses resolve early branches in the tree of life of modern birds," *Science* 346, no. 6215 (2014): 1320–31; G. Zhang et al., "Comparative genomics reveals insights into avian genome evolution and adaptation," *Science* 346, no. 6215 (2014): 1311–19.

44   **whereas in birds:** Oddly enough, the downy woodpecker is the one exception to this

rule; it has held on to 22 percent of its repeat elements. G. Zhang et al., "Comparative genomics reveals insights into avian genome evolution and adaptation," *Science* 346, no. 6215 (2014): 1311–19.

44 **Huxley, Darwin's bulldog**: H. G. Wells quoted in John Carey, *Eyewitness to Science* (Cambridge, MA: Harvard University Press, 1995), 139.

45 **but he saw birdlike traits**: P. Dodson, "Origin of birds: the final solution?" *Amer Zool* 40, no. 4 (2000): 504–12.

45 **Indeed, Huxley wrote**: T. H. Huxley, "Further evidence of the affinity between the dinosaurian reptiles and birds," *Proc Geol Soc Lond* (1870): 2612–31.

45 **In fact, says paleontologist**: Stephen Brusatte, personal communication, May 5, 2015.

45 **During the early Cretaceous**: M. J. Benton et al., "The remarkable fossils from the Early Cretaceous Jehol Biota of China and how they have changed our knowledge of Mesozoic life," *Proc Geol Assoc* 119 (2008): 209–28.

45 **When I was at one fossil site**: J. Ackerman, "Dinosaurs take wing: The origin of birds," *National Geographic* (July 1998): 74–99.

46 **The creature was a theropod**: Q. Ji et al., "Two feathered dinosaurs from northeastern China," *Nature* 393 (1998): 753–61; P. J. Chen, "An exceptionally well-preserved theropod dinosaur from the Yixian formation of China," *Nature* 391 (1998): 147–52, doi: 10.1038/34356; P. J. Currie and P. J. Chen, "Anatomy of *Sinosauropteryx prima* from Liaoning, northeastern China," *Can J Earth Sci* 38 (2001): 1705–27.

46 **One group of feathered dinosaurs**: According to Michael Benton of the University of Bristol, who commented on the research, "The crucial driver may have been a move to the trees, perhaps to escape from predators or to exploit new food resources. Tree living requires small bodies, enlarged eyes (the better to avoid collisions when leaping from branch to branch), and enlarged brains (to cope with diverse arboreal habitats). . . . These physical changes "are reminiscent of those that later occurred in our own clade, the primates, also interpreted as driven by tree living." See M. J. Benton, "How birds became birds," *Science* 345, no. 6196 (2014): 509.

46 **Dinosaurs gave rise to chickadees**: A. H. Turner, "A basal dromaeosaurid and size evolution preceding avian flight," *Science* 317, no. 5843 (2007): 1378–81; M. S. Y. Lee et al., "Sustained miniaturization and anatomical innovation in the dinosaurian ancestors of birds," *Science* 345, no. 6196 (2014): 562–66.

46 **More than 200 million years ago**: R. B. J. Benson et al., "Rates of dinosaur body mass evolution indicate 170 million years of sustained ecological innovation on the avian stem lineage," *PLoS Biol* 12, no. 5: e1001853, doi:10.1371/journal.pbio.1001853 (2014).

46 **Over a period of 50 million years**: M. S. Y. Lee et al., "Sustained miniaturization and anatomical innovation in the dinosaurian ancestors of birds," *Science* 345, no. 6196 (2014): 562–66.

47 **They developed new adaptations**: S. Brusatte et al., "Gradual assembly of avian body plan culminated in rapid rates of evolution across the dinosaur-bird transition," *Curr Biol* 24, no. 20 (2014): 2386–92.

47   **The dinosaurs that gave rise:** A. Balanoff et al., "Evolutionary origins of the avian brain," *Nature* 501 (2013): 93–96.

47   **When an international group:** B.-A. S. Bhullar et al., "Birds have paedomorphic dinosaur skulls," *Nature* 487 (2012): 223–26.

47   **"During juvenile to adult":** Arkhat Abzhanov, personal communication, January 25, 2015; A. Abzhanov quoted in University of Texas, Austin, press release, "Evolution of birds is result of a drastic change in how dinosaurs developed," May 30, 2012.

48   **Why do brood parasites:** J. R. Corfield et al., "Brain size and morphology of the brood-parasitic and cerophagous honeyguides (Aves: Piciformes)," *Brain Behav Evol* (February 2012), doi:10.1159/000348834; Louis Lefebvre, interview, February 2012.

49   **Brain size is also correlated:** A. N. Iwaniuk and J. E. Nelson, "Developmental differences are correlated with relative brain size in birds: A comparative analysis," *Can J Zool* 81 (2003): 1913–28.

50   **Pectoral sandpipers, for instance:** J. A. Lesku et al., "Adaptive sleep loss in polygynous pectoral sandpipers," *Science* 337 (2012): 1654–58.

50   **Birds experience the same cycles:** J. A. Lesku and N. C. Rattenborg, "Avian sleep," *Curr Biol* 24, no. 1 (2014): R12–R14.

51   **Likewise, studies show:** M. F. Scriba et al., "Linking melanism to brain development: Expression of a melanism-related gene in barn owl feather follicles covaries with sleep ontogeny," *Front Zool* 10 (2013): 42.

51   **A team of international researchers:** J. A. Lesku et al., "Local sleep homeostasis in the avian brain: Convergence of sleep function in mammals and birds?" *Proc R Soc B* 278 (2011): 2419–28.

51   **That both humans and birds:** Niels Rattenborg, personal communication, February 10, 2015.

53   **It may not pay to spend:** D. Sol, quoted in Autonomous University of Barcelona materials, http://www.alphagalileo.org/ViewItem.aspx?ItemId=74774&CultureCode =en.

53   **Vladimir Pravosudov of the University of Nevada:** T. C. Roth and V. V. Pravosudov, "Tough times call for bigger brains," *Commun Integ Biol* 2, no. 3 (May 2009): 236–38; V. V. Pravosudov and N. S. Clayton, "A test of the adaptive specialization hypothesis: Population differences in caching, memory, and the hippocampus in black-capped chickadees (*Poecile atricapilla*)," *Behav Neurosci* 116, no. 4 (2002): 515–22.

53   **Mountain chickadees that live:** C. A. Freas et al., "Elevation-related differences in memory and the hippocampus in mountain chickadees, *Poecile gambeli*," *Anim Behav* 84, no. 1 (2012): 121–27.

53   **(They are also better problem solvers):** V. V. Pravosudov, "Cognitive ecology of food-hoarding: The evolution of spatial memory and the hippocampus," *Ann Rev Ecol Evol Syst* 44 (2013): 18.1–18.2.

53   **Recovering caches is not:** Pravosudov suspects that the number of neurons in the hippocampus of these different populations of chickadees is inherited, "produced by

natural selection acting on memory rather than by individuals adjusting to changing conditions," he says. Vladimir Pravosudov, personal communication, January 23, 2015; V. V. Pravosudov et al., "Environmental influences on spatial memory and the hippocampus in food-caching chickadees," *Comp Cog and Beh Rev* (in press, 2015).

53 **The reason for this neurogenesis:** A. Barnea and V. V. Pravosudov, "Birds as a model to study adult neurogenesis: Bridging evolutionary, comparative and neuroethological approaches," *Eur J Neuroscience* 34 (2011): 884–907.

53 **It may allow the brain to recruit:** One hypothesis suggests that it provides a "neurogenic reserve," allowing the brain to remain flexible and to recruit new neurons when it's required to learn new information. Another theory proposes that these new neurons help to avoid "catastrophic interference" between old and new memories when the brain is learning something new. G. Kempermann, "The neurogenic reserve hypothesis: What is adult hippocampal neurogenesis good for?" *Trends Neurosci* 31 (2008): 163–69; L. Wiskott et al., "A functional hypothesis for adult neurogenesis: Avoidance of catastrophic interference in the dentate gyrus," *Hippocampus* 16 (2006): 329–43; W. Deng et al., "New neurons and new memories: How does adult hippocampal neurogenesis affect learning and memory?" *Nat Rev Neurosci* 11 (2010): 339–50.

53 **The "interference avoidance":** C. D. Clelland et al., "A functional role for adult hippocampal neurogenesis in spatial pattern separation," *Science* 325 (2009): 210–13.

53 **Pravosudov has shown:** T. C. Roth and V. V. Pravosudov, "Tough times call for bigger brains," *Commun Integ Biol* 2, no. 3 (May 2009): 236–38.

54 **The conventional notion:** S. Herculano-Houzel, "Neuronal scaling rules for primate brains: The primate advantage," *Prog Brain Res* 195 (2012): 325–40.

54 **In 2014, Brazilian:** S. Olkowicz et al., "Complex brains for complex cognition—neuronal scaling rules for bird brains" (poster presentation at the Society for Neuroscience annual meeting, Washington, D.C., November 15–19, 2014).

54 **The brains of birds:** Suzana Herculano-Houzel, personal communication, January 14, 2015.

54 **Herculano-Houzel has shown:** S. Herculano-Houzel et al., "The elephant brain in numbers," *Front Neuroanat* 8 (2014): 46, doi: 10.3389/fnana.2014.00046.

55 **"Birds were viewed as lovely":** H. Karten, quoted in S. LaFee, "Our brains are more like birds' than we thought," 2010, http://ucsdnews.ucsd.edu/archive/newsrel/health/07-02avianbrain.asp.

55 **This anatomical dissing:** Avian Brain Nomenclature Consortium, "Avian brains and a new understanding of vertebrate brain evolution," *Nat Rev Neurosci* 6, no. 2 (2005): 151–59; T. Shimizu, "Why can birds be so smart? Background, significance, and implications of the revised view of the avian brain," *Comp Cog Beh Rev* 4 (2009): 103–15.

56 **Instead of a layered:** As Peter Marler wrote, "Widespread presumption that surface cortical area is a direct correlate of intelligence prepared us for expectation that the smooth-surfaced avian brain will be found ill-designed to support high-level intellectual accomplishments." P. Marler, "Social cognition," in *Curr Orni* 13 (1996): 1–32.

56　**But beginning in the late 1960s:** H. J. Karten in *Comparative and Evolutionary Aspects of the Vertebrate Central Nervous System*, ed. J. Pertras, *Ann NY Acad Sci* 167 (1969): 164–79; H. J. Karten and W. A. Hodos, *A Stereotaxic Atlas of the Brain of the Pigeon* (Baltimore: Johns Hopkins University Press, 1967).

57　**What they found turned Edinger's:** Avian Brain Nomenclature Consortium, "Avian brains and a new understanding of vertebrate brain evolution," *Nat Rev Neurosci* 6, no. 2 (2005): 151–59.

57　**a pigeon's exceptional ability:** R. J. Herrnstein and D. H. Loveland, "Complex visual concept in the pigeon," *Science* 146 (1964): 549–51.

57　**An international group of twenty-nine:** Avian Brain Nomenclature Consortium, "Avian brains and a new understanding of vertebrate brain evolution," 151–59.

58　**"About 75 percent of our forebrain is cortex":** Erich Jarvis, interview, March 23, 2012.

58　**Irene Pepperberg uses a computer analogy:** I. M. Pepperberg, *The Alex Studies* (Boston: Harvard University Press, 1999), 9.

59　**To find out, Andreas Nieder:** L. Veit et al., "Neuronal correlates of visual working memory in the corvid endbrain, *J Neurosci* 34, no. 23 (2014): 7778–86.

59　**In humans and birds:** O. Güntürkün, "The convergent evolution of neural substrates for cognition," *Psychol Res* 76 (2012): 212–19.

60　**(Even the way the wild ibis):** B. Voelkl et al., "Matching times of leading and following suggest cooperation through direct reciprocity during V-formation flight in ibis," *PNAS* 112, no. 7 (2015): 2115–20.

Three: Boffins: *Technical Wizardry*

63　**Like 007, Blue:** Sources for general information on New Caledonian crows include my interviews with Alex Taylor, May 2014; and A. H. Taylor, "Corvid cognition," *Wiley Interdiscip Rev Cogn Sci* 5, no. 3 (2014): 361–72.

63　**In the wild, these crows:** L. A. Bluff et al., "Tool use by wild New Caledonian crows *Corvus moneduloides* at natural foraging sites," *Proc R Soc B* 277, no. 1686 (2010): 1377–85.

64　**they know a good tool:** B. C. Klump et al., "Context-dependent 'safekeeping' of foraging tools in New Caledonian crows," *Proc R Soc B* 282 (2015): 20150278.

64　**A list of the tools we have invented:** A. H. Taylor and R. D. Gray, "Is there a link between the crafting of tools and the evolution of cognition?" *Wiley Interdiscip Rev Cogn Sci* 5, no. 6 (2014): 693–703.

65　**The notion of tool use:** The following information on animal tool use is from R. W. Shumaker et al., *Animal Tool Behavior* (Baltimore: Johns Hopkins University Press, 2011).

65　**A female digger wasp:** H. J. Brockmann, "Tool use in digger wasps (*Hymenoptera: Sphecinae*)," *Psyche* 92 (1985): 309–30.

65 Still, tool use is exceedingly rare: D. Biro et al., "Tool use as adaptation," *Phil Trans R Soc Lond B* 368, no. 1630 (2013): 20120408.

65 Especially when you look at the catalog: E. Meulman and C. P. van Schaik, "Orangutan tool use and the evolution of technology," in ed. C. M. Sanz et al., *Tool Use in Animals: Cognition and Ecology* (New York: Cambridge University Press, 2013), 176.

65 Or a chimp's clever constructions: C. Boesch, "Ecology and cognition of tool use in chimpanzees," in C. M. Sanz et al., eds., *Tool Use in Animals: Cognition and Ecology*, 21–47.

65 While they may not make and use: W. C. McGrew, "Is primate tool use special? Chimpanzee and New Caledonian crow compared," *Philos Trans R Soc Lond B* 368 (2013): 20120422.

65 They make them the proper length: J. Chappell and A. Kacelnik, "Tool selectivity in a non-primate, the New Caledonian crow (*Corvus moneduloides*)," *Anim Cogn* 5 (2002): 71–78; J. Chappell and A. Kacelnik, "Selection of tool diameter by New Caledonian crows *Corvus moneduloides*," *Anim Cogn* 7 (2004): 121–27.

65 They use tools in sequence: J. H. Wimpenny et al., "Cognitive processes associated with sequential tool use in New Caledonian crows," *PLoS ONE* 4, no. 8 (2009): e6471, doi:10.1371/journal.pone.0006471.

66 Alex Taylor has taken: Quotes from Taylor that follow are from interviews in May 2014.

67 Not only do New Caledonian crows: K. D. Tanaka et al., "Gourmand New Caledonian crows munch rare escargots by dropping numerous broken shells of a rare endemic snail *Placostylus fibratus*, a species rated as vulnerable, were scattered around rocky beds of dry creeks in rainforest of New Caledonia," *J Ethol* 31 (2013): 341–44.

67 The vampire finch of the Galápagos: P. R. Grant, *Ecology and Evolution of Darwin's Finches* (Princeton, NJ: Princeton University Press, 1986), 393.

67 The black-breasted buzzard: R. W. Shumaker et al., *Animal Tool Behavior* (Baltimore: Johns Hopkins University Press, 2011), 38.

67 Carrion crows use passing cars: Y. Nihei, "Variations of behavior of carrion crows *Corvus corone* using automobiles as nutcrackers," *Jpn J Ornithol* 44 (1995): 21–35.

67 A wanderer in the pages: R. W. Shumaker et al., *Animal Tool Behavior* (Baltimore: Johns Hopkins University Press, 2011), 35–58.

68 for instance, a white stork: J. Rekasi, "Über die Nahrung des Weisstorchs (*Ciconia ciconia*) in der Batschka (SüdUngarn)," *Ornith Mit* 32 (1980): 154–55, cited in L. Lefebvre et al., "Tools and brains in birds," *Behaviour* 139 (2002): 939–73.

68 African grey parrots bailing: I. M. Pepperberg and H. A. Shive, "Simultaneous development of vocal and physical object combinations by a grey parrot (*Psittacus erithacus*): bottle caps, lids, and labels," *J Comp Psychol* 115 (2001): 376–84.

68 an American crow ferrying: P. D. Cole, "The ontogenesis of innovative tool use in an American crow (*Corvus brachyrhynchos*)" (PhD thesis, Dalhousie University, 2004).

68 a Gila woodpecker fashioning: L. Lefebvre, "Feeding innovations and forebrain size

in birds" (AAAS presentation, February 21, 2005, part of the symposium "Mind, Brain and Behavior").

68 **a blue jay using:** T. Eisner, "'Anting' in blue jays: Evidence in support of a food-preparatory function," *Chemoecology* 18, no. 4 (December 2008): 197–203.

68 **One American crow in Stillwater:** C. Caffrey, "Goal-directed use of objects by American crows," *Wilson Bulletin* 113, no. 1 (2001): 114–15.

68 **A pair of ravens in Oregon:** S. W. Janes et al., "The apparent use of rocks by a raven in nest defense," *Condor* 78 (1976): 409.

68 **Other notable bird uses:** R. W. Shumaker et al., *Animal Tool Behavior* (Baltimore: Johns Hopkins University Press, 2011), 35–58.

68 **as drumsticks by black palm cockatoos:** S. Taylor, *John Gould's Extinct and Endangered Birds of Australia* (Canberra: National Library of Australia, 2012), 130.

69 **Early one April morning:** R. P. Balda, "Corvids in combat: With a weapon?" *Wilson J Ornithol* 119, no. 1 (2007): 100.

69 **Among the handful of birds:** S. Tebbich, "Tool-use in the woodpecker finch *Cactospiʐa pallida*: Ontogeny and ecological relevance" (PhD thesis, University of Vienna, 2000). According to Gavin Hunt, the other species that use tools regularly are Egyptian vultures, black-breasted buzzards, brown-headed nuthatches, and black palm cockatoos. Gavin Hunt, personal communication, January 2015.

70 **Sabine Tebbich, a behavioral biologist:** S. Tebbich et al., "The ecology of tool-use in the woodpecker finch (*Cactospiʐa pallida*)," *Ecol Lett* 5 (2002): 656–64.

70 **In the first experimental study:** S. Tebbich, "Do woodpecker finches acquire tool-use by social learning?," *Proc R Soc B* 268 (2001): 1–5.

70 **two of them documented:** G. Merlen and G. Davis-Merlen, "Whish: More than a tool-using finch," *Noticias de Galápagos* 61 (2000): 2–9.

71 **Recently Tebbich and:** S. Tebbich et al., "Use of a barbed tool by an adult and a juvenile woodpecker finch (*Cactospiʐa pallida*)," *Behav Process* 89, no. 2 (2012): 166–71.

71 **Take Goffin's cockatoos:** A. M. I. Auersperg et al., "Explorative learning and functional inferences on a five-step means-means-end problem in Goffin's cockatoos (*Cacatua goffini*)," *PLoS ONE* 8, no. 7 (2013): e68979.

71 **But Alice Auersperg and:** A. M. I. Auersperg et al., "Spontaneous innovation in tool manufacture and use in a Goffin's cockatoo," *Curr Biol* 22, no. 21 (2012): R903–R904.

71 **A few years ago, Christian Rutz:** L. A. Bluff et al., "Tool use by wild New Caledonian crows *Corvus moneduloides* at natural foraging sites," *Proc R Soc B* 277 (2010): 1377–85; C. Rutz et al., "Video cameras on wild birds," *Science* 318, no. 5851 (2007): 765.

72 **The crows repeatedly poke:** C. Rutz and J. J. H. St Clair, "The evolutionary origins and ecological context of tool use in New Caledonian crows," *Behav Proc* 89 (2012): 153–65.

72 **Rutz and his colleagues:** Ibid., 156.

72 **And not even these hotshot:** G. R. Hunt and R. D. Gray, "The crafting of hook tools by wild New Caledonian crows," *Proc R Soc B* (suppl.) 271 (2004): S88–S90.

72 **The pandanus tools:** G. R. Hunt, "Manufacture and use of hook-tools by New Caledonian crows," *Nature* 379 (1996): 249–51; G. R. Hunt and R. D. Gray, "Species-wide
manufacture of stick-type tools by New Caledonian crows," *Emu* 102 (2002): 349–53;
G. R. Hunt and R. D. Gray, "Diversification and cumulative evolution in tool manufacture by New Caledonian crows," *Proc R Soc B* 270 (2003): 867–74; G. R. Hunt and
R. D. Gray, "The crafting of hook tools by wild New Caledonian crows," *Proc R Soc B*
(suppl.) 271 (2004): S88–S90; G. R. Hunt and R. D. Gray, "Direct observations of
pandanus-tool manufacture and use by a New Caledonian crow (*Corvus moneduloides*)," *Anim Cogn* 7 (2004): 114–20; C. Rutz and J. J. H. St Clair, "The evolutionary
origins and ecological context of tool use in New Caledonian crows," *Behav Processes*
89, no. 2 (2012): 153–65.

72 **It takes many complex moves:** G. R. Hunt, "Manufacture and use of hook-tools by
New Caledonian crows," *Nature* 379 (1996): 249–51; G. R. Hunt and R. D. Gray, "Di-
rect observations of pandanus-tool manufacture and use by a New Caledonian crow
(*Corvus moneduloides*)."

73 **One remarkable feature:** J. C. Holzhaider et al., "Social learning in New Caledonian
crows," *Learn Behav* 38, no. 3 (2010): 206–19.

73 **In an island-wide survey:** G. R. Hunt and R. D. Gray, "Diversification and cumulative evolution in tool manufacture by New Caledonian crows."

73 **Faithful transmission of local:** L. G. Dean et al., "Identification of the social
and cognitive processes underlying human cumulative culture," *Science* 335 (2012):
1114–18.

73 **Moreover, in Hunt's view:** Gavin Hunt, personal communication, January 2015;
G. R. Hunt, "New Caledonian crows' (*Corvus moneduloides*) pandanus tool designs: Diversification or independent invention?" *Wilson J Ornithol* 126, no. 1 (2014): 133–39;
G. R. Hunt and R. D. Gray, "Diversification and cumulative evolution in tool manufacture by New Caledonian crows."

74 **Christian Rutz argues:** C. Rutz and J. J. H. St Clair, "The evolutionary origins and
ecological context of tool use in New Caledonian crows."

74 **In a suite of experiments:** J. J. H. St Clair and C. Rutz, "New Caledonian crows at-
tend to multiple functional properties of complex tools," *Phil Trans R Soc Lond B* 368,
no. 1630 (2013): 20120415.

74 **Why is it that:** The following discussion of the unique characteristics and possible
evolutionary origins of the crows' tool use draws on the brilliant review paper by
C. Rutz and J. J. H. St Clair, "The evolutionary origins and ecological context of tool
use in New Caledonian crows."

74 **A remote sprig of land:** Information on New Caledonia from Conservation International Web site page, http://sp10.conservation.org/where/asia-pacific/pacific_
islands/new_caledonia/Pages/overview.aspx; C. Rutz and J. J. H. St Clair, "The evolutionary origins and ecological context of tool use in New Caledonian crows,"
153–65.

75 **It has roughly the land area of New Jersey:** http://newcaledoniaplants.com/.

75 **When Cook approached:** http://newcaledoniaplants.com/plant-catalog/arau carians/.

75 **In the gloom below:** M. G. Fain and P. Houde, "Parallel radiations in the primary clades of birds," *Evolution* 58 (2004): 2558–73.

75 **But the island remains a diversity:** A. Gasc et al., "Biodiversity sampling using a global acoustic approach: Contrasting sites with microendemics in New Caledonia," *PLoS ONE* 8, no. 5 (2013): e65311.

75 **There are some 3,200 species:** There are about 3,270 plant species recorded on the islands, 74 percent of which are endemic (roughly 2,430 species): http://www.cepf .net/resources/hotspots/Asia-Pacific/Pages/New-Caledonia.aspx.

75 **It is also an ark of colossal creatures:** http://sp10.conservation.org/where/asia -pacific/pacific_islands/new_caledonia/Pages/overview.aspx.

76 **According to Christian Rutz:** It's possible that some land remained above water and that crows lived on this small island refuge. This scenario might explain the presence of the kagu, too. See C. Rutz and J. J. H. St Clair, "The evolutionary origins and ecological context of tool use in New Caledonian crows."

77 **For animals clever enough:** Following information on ecology of New Caledonian crow tool use from C. Rutz and J. J. H. St Clair, "The evolutionary origins and ecological context of tool use in New Caledonian crows"; C. Rutz et al., "The ecological significance of tool use in New Caledonian crows," *Science* 329, no. 5998 (2010): 1523–26.

77 **The grubs are heavy:** C. Rutz et al., "The ecological significance of tool use in New Caledonian crows"; C. Rutz et al., "Video cameras on wild birds," *Science* 318, no. 5851 (2007): 765.

77 **One consequence of such modest threat:** C. Rutz and J. J. H. St Clair, "The evolutionary origins and ecological context of tool use in New Caledonian crows."

77 **One experiment showed:** B. Kenward et al., "Tool manufacture by naïve juvenile crows," *Nature* 433 (2005): 121; B. Kenward et al., "Development of tool use in New Caledonian crows: Inherited action patterns and social influences," *Anim Behav* 72 (2006): 1329–43.

78 **As a PhD student studying:** J. C. Holzhaider et al., "Social learning in New Caledonian crows."

78 **In his lecture on the evolution:** The following description draws on J. C. Holzhaider et al., "Social learning in New Caledonian crows"; notes from personal communication with Jenny Holzhaider and from her 2011 radio interview on 95bFM at http://www .95bfm.co.nz/assets/sm/198489/3/RSL_8.02.11.mp3 ; as well as Russell Gray's fascinating analysis of Yellow-Yellow's learning process in his 2014 lecture "The evolution of cognition without miracles" (Nijmegen Lectures, January 27–29, 2014), video recording at: http://www.mpi.nl/events/nijmegen-lectures-2014/lecture-videos.

78 **This may help to explain the existence:** According to Hunt, Holzhaider et al., "We believe the strongest evidence of human-like cumulative technological evolution in a

nonhuman is provided by NC crows manufacture of tools from the leaves of Pandanus species trees." J. C. Holzhaider et al., "Social learning in New Caledonian crows."

78 **By observing and using tools:** R. Gray, "The evolution of cognition without miracles."

79 **According to the Auckland team:** G. R. Hunt, J. C. Holzhaider, and R. D. Gray, "Prolonged parental feeding in tool-using New Caledonian crows," *Ethology* 188 (2012): 1–8.

80 **A hidden trove of rich food:** C. Rutz and J. J. H. St Clair, "The evolutionary origins and ecological context of tool use in New Caledonian crows."

81 **They are also positioned more:** J. Troscianko et al., "Extreme binocular vision and a straight bill facilitate tool use in New Caledonian crows," *Nat Comm* 3 (2012): 1110.

81 **New research by Alex Kacelnik:** A. Martinho et al., "Monocular tool control, eye dominance, and laterality in New Caledonian crows," *Curr Biol* 24, no. 24 (2014): 2930–34.

81 **As Kacelnik puts it:** A. Kacelnik, quoted in "Why tool-wielding crows are left- or right-beaked," *Cell Press* 4 (December 2014), http://phys.org/news/2014-12-tool -wielding-crows-left-right-beaked.htm.

81 **The bird's beak, for its part:** J. Troscianko et al., "Extreme binocular vision and a straight bill facilitate tool use in New Caledonian crows," *Nat Comm* 3 (2012): 1110.

81 **It's not clear which came first:** D. Biro et al., "Tool use as adaptation," *Phil Trans R Soc Lond B* 368, no. 1630 (2013): 20120408.

81 **In any case, say scientists:** J. Troscianko et al., "Extreme binocular vision and a straight bill facilitate tool use in New Caledonian crows," *Nat Comm* 3 (2012): 1110.

82 **As Gavin Hunt points out:** Gavin Hunt, personal communication, January 21, 2015.

82 **Research suggests that there are small differences:** R. Gray, "The evolution of cognition without miracles."

82 **One study shows that the New Caledonian:** J. Cnotka et al., "Extraordinary large brains in tool-using New Caledonian crows (*Corvus moneduloides*)," *Neurosci Lett* 433 (2008): 241–45. Some scientists are skeptical of the method and analysis of this study. "The published evidence for tool-related neurological adaptations in NC crows is weak at best," write Christian Rutz and J. J. H. St Clair. See Rutz and St Clair, "The evolutionary origins and ecological context of tool use in New Caledoninan crows."

82 **There is some ballooning:** J. Mehlhorn, "Tool-making New Caledonian crows have large associative brain areas," *Brain Behav Evolut* 75 (2010): 63–70.

82 **Moreover, as Russell Gray points out:** R. Gray, "The evolution of cognition without miracles"; F. S. Medina et al., "Perineuronal satellite neuroglia in the telencephalon of New Caledonian crows and other Passeriformes: Evidence of satellite glial cells in the central nervous system of healthy birds?" *Peer J* 1 (2013): e110.

82 **In sum, the brains:** R. Gray, "The evolution of cognition without miracles."

83 **They're less interested in trumpeting:** The following is from an interview with Alex

Taylor; and A. Taylor, "Corvid cognition," *WIREs Cogn Sci* (2014), doi:10.1002/wcs.1286.

83 **The signatures of cognition evident:** R. Gray, "The evolution of cognition without miracles."

83 **According to Russell Gray:** A. H. Taylor et al., "Spontaneous metatool use by New Caledonian crows," *Curr Biol* 17 (2007): 1504–7; R. Gray, "The evolution of cognition without miracles."

84 **If he was using mental scenario:** A. H. Taylor, "Corvid cognition," *WIREs Cogn Sci* (2014), doi:10.1002/wcs.1286.

84 **The actions of 007 could be:** Alex Taylor, personal communication, January 7, 2015.

84 **After christening more than:** Some scientists, including Christian Rutz, believe it is better not to name study subjects, "as this may well affect the way experimenters observe/score experimental trials, and interpret evidence," says Rutz. Christian Rutz, personal communication, July 30, 2015.

85 **To see whether this is so:** A. H. Taylor et al., "An end to insight? New Caledonian crows can spontaneously solve problems without planning their actions," *Proc R Soc B* 279, no. 1749 (2012): 4977–81; Alex Taylor, interview.

86 **If this had been an example of insight:** But see A. M. Seed and N. J. Boogert, "Animal cognition: An end to insight?" *Curr Biol* 23, no. 2 (2013): R67–R69.

86 **The Auckland team keeps the birds:** Christian Rutz believes that moving birds between localities in New Caledonia is very risky. "If there are learned components to these birds' tool behaviour, exposing crows to techniques they are not familiar with may alter local 'traditions' or 'cultures'. Our team always tests crows *in situ* (that is, where they were trapped), to avoid such inadvertent 'contamination' of populations." Christian Rutz, personal communication, July 30, 2015.

86 **As it turns out:** S. A. Jelbert et al., "Using the Aesop's fable paradigm to investigate causal understanding of water displacement by New Caledonian crows," *PloS One* 9, no. 3 (2014): 1–9.

87 **Lately, Taylor and Gray:** A. H. Taylor et al., "New Caledonian crows reason about hidden causal agents," *PNAS* 109, no. 40 (2012): 16389–91.

87 **"We're constantly making inferences":** R. Gray, "The evolution of cognition without miracles."

87 **An infant only seven to ten months:** R. Saxe et al., "Knowing who dunnit: Infants identify the causal agent in an unseen causal interaction," *Develop Psych* 43, no. 1 (2007): 149–58; R. Saxe et al., "Secret agents: Inferences about hidden causes by 10- and 12-month-old infants," *Psychol Sci* 16, no. 12 (2005): 995–1001.

87 **As Gray points out:** R. Gray, "The evolution of cognition without miracles."

89 **The difference in the crows' behavior:** Critics of this study have suggested that the crows might not be exercising causal reasoning but rather just associating the poking of the stick with a human presence inside the hide. See N. J. Boogert et al., "Do

crows reason about causes or agents? The devil is in the controls," *PNAS* 110, no. 4 (2013): E273. "Yes, there's an association there," Taylor agrees. "If they see the stick move, then a human's going to come out of the hide. But this explanation doesn't explain why the crows aren't scared after the human leaves. The association account suggests that the crows are suicidal, so dumb that they're happy to put their heads exactly where the stick will emerge." See A. H. Taylor et al., "Reply to Boogert et al: The devil is unlikely to be in association or distraction," *PNAS* 110, no. 4 (2013): E274.

90 **But crows fail at the task:** A. H. Taylor et al., "Of babies and birds: Complex tool behaviours are not sufficient for the evolution of the ability to create a novel causal intervention," *Proc R Soc B* 281, no. 1787 (2014): 1–6.

90 **Nathan Emery, senior lecturer:** N. J. Emery and N. S. Clayton, "Do birds have the capacity for fun?" *Curr Biol* 25, no. 1 (2015): R16–R19.

91 **That is, play can be:** W. H. Thorpe in M. Ficken, "Avian Play," *Auk* 94 (1977): 574.

91 **According to zoologist Millicent Ficken:** M. Ficken, "Avian Play," *Auk* 94 (1977): 573–82.

91 **"Nestor was a legendary Greek hero":** A. F. Gotch, *Latin Names Explained* (New York: Facts on File, 1995), 286.

92 **Two scientists, Judy Diamond and Alan Bond:** J. Diamond and A. B. Bond, *Kea: Bird of Paradox* (Berkeley and Los Angeles: University of California Press, 1999), 76.

92 **The kea's playfulness:** Ibid., 99.

92 **A few years ago, the New Zealand:** M. Miller, "Parrot Steals $1100 from Unsuspecting Tourist," *Sunday Morning Herald*, February 4, 2013, http://www.traveller.com .au/parrot-steals-1100-from-unsuspecting-tourist-2dtc2.

93 **Two young white-necked ravens:** R. Moreau and W. Moreau, "Do young birds play?" *Ibis* 86 (1944): 93–94.

93 **One clear sunny February morning:** M. Brazil, "Common raven *Corvus corax* at play; records from Japan," *Ornithol Sci* 1 (2002): 150–52.

93 **Alice Auersperg and an international team:** A. M. I. Auersperg et al., "Combinatory actions during object play in psittaciformes (*Diopsittaca nobilis*, *Pionites melanocephala*, *Cacatua goffini*) and corvids (*Corvus corax*, *C. monedula*, *C. moneduloides*)," *J Comp Psych* 129, no. 1 (2015): 62–71; A. M. I. Auersperg et al., "Unrewarded object combinations in captive parrots," *Anim Behav Cogn* 1, no. 4 (2014): 470–88.

94 **The Goffin's favored yellow toys:** Keas are also drawn to yellow objects, and they, too, have yellow stripes beneath their wings. A. M. I. Auersperg et al., "Unrewarded object combinations in captive parrots," *Anim Behav Cogn* 1, no. 4 (2014): 470–88.

95 **One serious question the Auckland team:** The following discussion draws on interviews with Alex Taylor; and C. Rutz and J. J. H. St Clair, "The ecological significance of tool use in New Caledonian crows."

95 **It's possible that life:** C. Rutz and J. J. H. St Clair, "The ecological significance of tool use in New Caledonian crows."

95 **(So rich a food source):** J. R. Beggs and P. R. Wilson, "Energetics of South Island kaka (*Nestor meridionalis*) feeding on the larvae of kanuka longhorn beetles (*Ochrocydus huttoni*)," *New Zealand J Ecol* 10 (1987): 143–47.

96 **Of course, as Gavin Hunt points out:** Gavin Hunt, interview, May 12, 2014.

97 **The primeval mountain forests:** http://newcaledoniaplants.com/plant-catalog/humid-forest-plants/.

Four: Twitter: *Social Savvy*

101 **We "rub and polish":** *Complete Essays of Montaigne*, trans. D. Frame (Stanford, CA: Stanford University Press, 1958), book 1, chapter 26, 112.

102 **rooks, for instance:** P. Green, "The communal crow," *BBC Wildlife* 14, no. 1 (1996), 30–34.

102 **Great tits (*Parus major*):** L. M. Aplin et al., "Social networks predict patch discovery in a wild population of songbirds," *Proc R Soc B* 279 (2012): 4199–205.

102 **Even chickens form complex:** T. Schjelderup-Ebbe, "Contributions to the social psychology of the domestic chicken," in *Social Hierarchy and Dominance*, ed. M. Schein (Stroudsburg, PA: Dowden, Hutchinson & Ross, 1975), 35–49. However, if the chickens are separated for a few weeks, they tend to forget their dominance relationships. See T. Schjelderup-Ebbe, "Social behavior in birds," in *Handbook of Social Dynamics of Hierarchy Formation*, ed. C. Murchison (Worcester, MA: Clark University Press, 1935), 947–72.

103 **The idea that a demanding social life:** N. Humphrey, "The social function of intellect," initially published in *Growing Points in Ethology*, ed. P. P. G. Bateson and R. A. Hinde (Cambridge: Cambridge University Press, 1976), 303–17. The idea first originated with M. R. A. Chance and A. P. Mead, "Social behavior and primate evolution," *Symp Soc Exp Biol* 7 (1953): 395–439; and A. Jolly, "Lemur social behavior and primate intelligence," *Science* 153 (1966): 501–6.

104 **Magpies recognize their own:** H. Prior et al., "Mirror-induced behavior in the magpie (*Pica pica*): Evidence of self-recognition," *PLoS Biol* 6, no. 8 (2008): e202.

104 **In the wild, these birds:** T. Juniper and M. Parr, *Parrots: A Guide to Parrots of the World* (New Haven, CT: Yale University Press, 1998), 22.

104 **They're rarely alone:** African grey parrots kept alone in cages sometimes show signs of severe stress, pulling at their own feathers or screaming. Scientists lately found that social isolation actually damages the birds' chromosomes, shortening their telomeres, those little caps on chromosomes that are likened to the plastic tips on shoelaces because they keep the ends of chromosomes from fraying. See C. S. Davis, "Parrot psychology and behavior problems," *Vet Clin North Am Small Anim Pract* 21 (1991): 1281–88; D. Aydinonat et al., "Social isolation shortens telomeres in African grey parrots (*Psittacus erithacus erithacus*)," *PLoS ONE* 9, no. 4 (2014): e93839.

104 **They also understand the benefits:** F. Peron et al., "Human–grey parrot (*Psittacus erithacus*) reciprocity," *Anim Cogn* (2014), doi:10.1007/s10071-014-0726-3.

104 **But in recent years, tales:** "Birds That Bring Gifts and Do the Gardening," *BBC News Magazine*, March 10, 2015, http://www.bbc.com/news/magazine-31795681.

105 **even a candy heart:** J. Marzluff and T. Angell, *Gifts of the Crow* (New York: Free Press, 2012), 108.

105 **In 2015, a story:** K. Sewall, "The Girl Who Gets Gifts from Birds," *BBC News Magazine*, February 25, 2015, http://www.bbc.com/news/magazine-31604026.

105 **"Leaving gifts suggests":** J. Marzluff and T. Angell, *Gifts of the Crow*, 114.

105 **Crows and ravens will balk:** C. A. F. Wascher and T. Bugnyar, "Behavioral responses to inequity in reward distribution and working effort in crows and ravens," *PLoS ONE* 8, no. 2 (2013): e56885.

105 **Corvids and cockatoos:** V. Dufour et al., "Corvids can decide if a future exchange is worth waiting for," *Biol Lett* 8, no. 2 (2012): 201–4.

105 **Alice Auersperg and her team:** A. M. I. Auersperg et al., "Goffin cockatoos wait for qualitative and quantitative gains but prefer 'better' to 'more,'" *Biol Lett* 9 (2013): 20121092.

106 **Young ravens belong:** T. Bugnyar, "Social cognition in ravens," *Comp Cogn Behav Rev* 8 (2013): 1–12.

106 **They pick special individuals:** O. N. Fraser and T. Bugnyar, "Do ravens show consolation? Responses to distressed other," *PLoS ONE* 5, no. 5 (2010): e10605.

106 **Thomas Bugnyar, a cognitive biologist:** M. Boeckle and T. Bugnyar, "Long-term memory for affiliates in ravens," *Curr Biol* 22 (2012): 801–6.

106 **Just ask Bernd Heinrich:** B. Heinrich, *Mind of the Raven* (New York: Harper Perennial, 2007), 176.

106 **Or John Marzluff:** J. M. Marzluff, "Lasting recognition of threatening people by wild American crows," *Anim Behav* 79 (2010): 699–707.

107 **The disgruntled crows:** John Marzluff, personal communication, February 10, 2015.

107 **In a brain-imaging study:** J. M. Marzluff et al., "Brain imaging reveals neuronal circuitry underlying the crow's perception of human faces," *PNAS* 109, no. 39 (2012): 15912–17.

107 **Pinyon jays:** G. C. Paz-y-Miño et al., "Pinyon jays use transitive inference to predict social dominance," *Nature* 430 (2004): 778.

108 **Ljerka Ostojić and her colleagues:** L. Ostojić et al., "Can male Eurasian jays disengage from their own current desire to feed the female what she wants?" *Biol Lett* 10 (2014): 20140042; L. Ostojić et al., "Evidence suggesting that desire–state attribution may govern food sharing in Eurasian jays," *PNAS* 110 (2013): 4123–28.

109 **"These experiments provide":** Ljerka Ostojić, personal communication, April 2015.

109 **"Attributing desires to others":** Ibid.

110 **Two scientists at the University of Pennsylvania:** R. M. Seyfarth and D. L. Cheney,

"Affiliation, empathy, and the origins of theory of mind," *PNAS* (suppl.) 110, no. 2 (2013): 10349–56.

110 **Rooks and ravens, for example:** T. Bugnyar and K. Kotrschal, "Scrounging tactics in free-ranging ravens," *Ethology* 108 (2002): 993–1009; P. Green, "The communal crow," *BBC Wildlife* 14, no. 1 (1996): 30–34.

111 **Some chickadees are bold, "fast":** L. M. Guillette et al., "Individual differences in learning speed, performance accuracy and exploratory behavior in black-capped chickadees," *Anim Cogn* 18, no. 1 (2015): 165–78.

111 **It also found that bolder birds:** L. M. Aplin et al., "Social networks predict patch discovery in a wild population of songbirds," *Proc R Soc B* 279 (2012): 4199–4205.

111 **"This is especially important":** Lucy Aplin, personal communication, March 10, 2015.

111 **The team also found that different species:** L. M. Aplin et al., "Social networks predict patch discovery in a wild population of songbirds"; D. R. Farine, "Interspecific social networks promote information transmission in wild songbirds," *Proc R Soc B* 282 (2015): 20142804; Lucy Aplin, personal communication, March 10, 2015.

111 **In Sweden and Finland:** J. T. Seppanen and J. T. Forsman, "Interspecific social learning: Novel preference can be acquired from a competing species," *Curr Biol* 17 (2007): 1248–52.

112 **To see how this social learning:** L. M. Aplin et al., "Experimentally induced innovations lead to persistent culture via conformity in wild birds," *Nature* 518, no. 7540 (2014): 538–41.

112 **A year later, birds remembered:** Lucy Aplin, personal communication, March 10, 2015.

113 **It is also, says Neeltje Boogert:** N. Boogert, "Milk bottle-raiding birds pass on thieving ways to their flock," *The Conversation*, December 4, 2014, https://theconversation.com/milk-bottle-raiding-birds-pass-on-thieving-ways-to-their-flock-34784.

113 **Female zebra finches learn:** J. P. Swaddle et al., "Socially transmitted mate preferences in a monogamous bird: A non-genetic mechanism of sexual selection," *Proc R Soc B* 272 (2005): 1053–58.

113 **One experiment showed that European blackbirds:** E. Curio et al., "Cultural transmission of enemy recognition: One function of mobbing," *Science* 202 (1978): 899.

113 **Young superb fairywrens:** W. E. Feeney and N. E. Langmore, "Social learning of a brood parasite by its host," *Biol Letters* 9 (2013): 20130443.

113 **A brilliant string of studies:** J. M. Marzluff, "Lasting recognition of threatening people by wild American crows," *Anim Behav* 79 (2010): 699–707.

114 **But growing evidence suggests:** T. M. Caro and M. D. Hauser, "Is there teaching in nonhuman animals?" *Q Rev Biol* 67 (1992): 151.

114 **Meerkats, for instance:** A. Thornton and K. McAuliffe, "Teaching in wild meerkats," *Science* 313 (2006): 227–29.

115 **Scientists have observed:** N. R. Franks and T. Richardson, "Teaching in tandem-running ants," *Nature* 439, no. 153 (2006), doi:10.1038/439153a.

115 **The birds live in small:** Amanda Ridley, personal communication, March 11, 2015.

115 **The dominant pair is monogamous:** M. J. Nelson-Flower et al., "Monogamous dominant pairs monopolize reproduction in the cooperatively breeding pied babbler," *Behav Ecol* (2011), doi:10.1093/beheco/arr018.

115 **In any group, 95 percent:** Ibid.

115 **Still, all the adults:** A. R. Ridley and N. J. Raihani, "Facultative response to a klepto-parasite by the cooperatively breeding pied babbler," *Behav Ecol* 18 (2007): 324–30; A. R. Ridley et al., "The cost of being alone: The fate of floaters in a population of cooperatively breeding pied babblers Turdoides bicolor," *J Avian Biol* 39 (2008): 389–92.

115 **If the pair does not produce:** "The re-occurrence of an extraordinary behaviour: A new kidnapping event in the population," *Pied & Arabian Babbler Research* (blog), November 2012, http://www.babbler-research.com/news.html. "The big news from Lizzy is that there is a new kidnapping event in the population! This is EXTREMELY interesting for us. Kidnapping is a rare behaviour, but entirely unexpected, and happens far more often than we would have ever imagined. This kidnapping fits the profile: CMF, a very small group who have failed to raise their own young for one and a half years (and are therefore at high risk of extinction), have stolen one of SHA group's very young fledglings. They are looking after it as if it was their own. We will continue to monitor this intriguing relationship between kidnapper and kidnapee [*sic*]."

116 **The sentinel perches in an open spot:** A. R. Ridley et al., "Is sentinel behaviour safe? An experimental investigation," *Anim Behav* 85, no. 1 (2012): 137–42.

116 **Small solitary birds:** A. R. Ridley et al., "The ecological benefits of interceptive eavesdropping," *Funct Ecol* 28, no. 1 (2013): 197–205.

116 **This allows the lone scimitarbills:** Ibid.

116 **Highly intelligent, accomplished mimics:** T. P. Flower, "Deceptive vocal mimicry by drongos," *Proc R Soc B* (2010), doi:10.1098/rspb.2010.1932.

116 **Ridley and her team recently:** T. P. Flower et al., "Deception by flexible alarm mimicry in an African bird," *Science* 344 (2014): 513–16.

116 **Ridley and her colleague:** N. J. Raihani and A. R. Ridley, "Adult vocalizations during provisioning: Offspring response and postfledging benefits in wild pied babblers," *Anim Behav* 74 (2007): 1303–9; N. J. Raihani and A. R. Ridley, "Experimental evidence for teaching in wild pied babblers," *Anim Behav* 75 (2008): 3–11. As Raihani and Ridley observe, to be classified as teaching, an interaction between two animals must do three things: "Teachers" must modify their behavior only in the presence of a naïve pupil. They must pay some sort of cost or at least gain nothing from modifying their behavior. And as a result of their modified behavior, the pupil must acquire knowledge or learn a skill faster than it would have otherwise.

117 **First, they're picky about whom:** A. M. Thompson and A. R. Ridley, "Do fledglings

choose wisely? An experimental investigation into social foraging behavior," *Behav Ecol Sociobiol* 67, no. 1 (2013): 69–78.

117 **Second, when they're hungry:** A. M. Thompson et al., "The influence of fledgling location on adult provisioning: A test of the blackmail hypothesis," *Proc R Soc B* 280 (2013): 20130558.

117 **It remains an open question:** J. A. Thornton and A. McAuliffe, "Cognitive consequences of cooperative breeding? A critical appraisal," *J Zool* 295 (2015): 12–22.

118 **"Teaching pied babblers":** Amanda Ridley, personal communication, April 7, 2015.

118 **What they haven't found:** G. Beauchamp and E. Fernandez-Juricic, "Is there a relationship between forebrain size and group size in birds?" *Evol Ecol Res* 6 (2004): 833–42.

118 **Indeed, when Oxford anthropologist and evolutionary psychologist Robin Dunbar:** R. Dunbar and S. Shultz, "Evolution in the social brain," *Science* 317 (2007): 1344–47.

118 **A clever computer simulation:** L. McNally et al., "Cooperation and the evolution of intelligence," *Proc R Soc B* (April 2012), doi:10.1098/rspb.2012.0206.

119 **However, when Dunbar and his colleagues looked:** S. Shultz and R. I. M. Dunbar, "Social bonds in birds are associated with brain size and contingent on the correlated evolution of life-history and increased parental investment," *Biol J Linn Soc* 100 (2010): 111–23.

119 **For birds, it seems, the quality of relationships:** "It is the qualitative nature (rather than the quantitative number) of relationships that imposes the cognitive burden." Ibid.

119 **The really demanding task:** N. J. Emery et al., "Cognitive adaptations to bonding in birds," *Philos Trans R Soc Lond B Biol Sci* 362 (2007): 489–505.

119 **About 80 percent of bird species live in socially monogamous pairs:** A. Cockburn, "Prevalence of different modes of parental care in birds," *Proc R Soc B* 273 (2006): 1375–83.

120 **According to cognitive biologist Nathan Emery:** N. J. Emery et al., "Cognitive adaptations to bonding in birds," *Philos Trans R Soc Lond B Biol Sci* 362 (2007): 489–505.

120 **Rook pairs, for instance:** N. S. Clayton and N. J. Emery, "The social life of corvids," *Curr Biol* 17, no. 16 (2007): R652–R656.

120 **Plain-tailed wrens:** E. Fortune et al., "Neural mechanisms for the coordination of duet singing in wrens," *Science* 334 (2011): 666–70.

120 **A male budgerigar:** M. Moravec et al., "'Virtual parrots' confirm mating preferences of female budgerigars," *Ethology* 116, no. 10 (2010): 961–71.

120 **After only a few days together, pair-bonded budgerigars:** A. G. Hile et al., "Male vocal imitation produces call convergence during pair bonding in budgerigars," *Anim Behav* 59 (2000): 1209–18.

121 **"It could also explain why parrot enthusiasts suggest":** Ibid.

121 **According to Goodson, the circuits in the brains:** L. A. O'Connell et al., "Evolution of a vertebrate social decision-making network," *Science* 336, no. 6085 (2012): 1154–57.

121 **The circuits are old:** J. L. Goodson and R. R. Thompson, "Nonapeptide mechanisms of social cognition, behavior and species-specific social systems," *Curr Opin Neurobiol* 20 (2010): 784–94.

121 **In birds, Goodson found that differences in social behavior:** J. L. Goodson, "Nonapeptides and the evolutionary patterning of social behavior," *Prog Brain Res* 170 (2008): 3–15.

122 **In the early 1990s, neuroendocrinologist Sue Carter:** C. S. Carter et al., "Oxytocin and social bonding," *Ann NY Acad Sci* 652 (1992): 204–11.

122 **New research shows that food sharing in chimps:** C. Crockford et al., "Urinary oxytocin and social bonding in related and unrelated wild chimpanzees," *Proc R Soc B* 280 (2013): 20122765.

122 **In humans, oxytocin has been shown to reduce anxiety:** M. Heinrichs et al., "Oxytocin, vasopressin, and human social behavior," *Front Neuroendocrin* 30 (2009): 548–57; K. MacDonald and T. M. MacDonald, "The peptide that binds: A systematic review of oxytocin and its prosocial effects in humans," *Harvard Rev Psychiat* 18, no. 1 (2010): 1–21.

122 **Recent studies have suggested, for instance, that a dose of oxytocin:** G.-J. Pepping and E. J. Timmermans, "Oxytocin and the biopsychology of performance in team sports," *Sci World J* (2012): 567363.

122 **It also may contribute to the strength of romantic bonds:** D. Scheele et al., "Oxytocin enhances brain reward system responses in men viewing the face of their female partner," *Proc Natl Acad Sci* 110, no. 5 (2013): 20308020313.

122 **The biologists discovered:** J. L. Goodson and M. A. Kingsbury, "Nonapeptides and the evolution of social group sizes in birds," *Front Neuroanat* 5 (2011): 13; J. L. Goodson et al., "Evolving nonapeptide mechanisms of gregariousness and social diversity in birds," *Horm Behav* 61 (2012): 239–50.

122 **On the other hand, birds that were given mesotocin:** J. L. Goodson et al., "Mesotocin and nonapeptide receptors promote songbird flocking behavior," *Science* 325 (2009): 862–66.

123 **(and known in one lab):** Neeltje Boogert, personal communication, April 7, 2015.

123 **When Goodson mapped the oxytocin-like receptors:** J. L. Goodson et al., "Mesotocin and nonapeptide receptors promote songbird flocking behavior," *Science* 325 (2009): 862–66.

123 **Curious about whether the oxytocin-like peptides:** J. D. Klatt and J. L. Goodson, "Oxytocin-like receptors mediate pair bonding in a socially monogamous songbird," *Proc R Soc B* 280, no. 1750 (2012): 20122396.

123 **In one study, psychologist Ruth Feldman:** R. Feldman, "Oxytocin and social affiliation in humans," *Horm Behav* 61 (2012): 380–91.

123 **However, as Marcy Kingsbury points out:** Marcy Kingsbury, personal communication, February 9, 2015; and see J. L. Goodson et al., "Oxytocin mechanisms of stress response and aggression in a territorial finch," *Physiol Behav* 141 (2015): 154–63. The authors write, "OT can promote negative behavior and perceptions, as is increasingly well documented in humans. For instance, intranasal OT administration reduces trust and cooperation in borderline personality patients, and promotes parochial altruism, ethnocentrism, and out-group derogation in healthy men."

124 **Indeed, some studies of human couples:** S. E. Taylor et al., "Are plasma oxytocin in women and plasma vasopressin in men biomarkers of distressed pair-bond relationships?" *Psychol Sci* 21 (2010): 3–7.

124 **According to Rhiannon West, a biologist:** R. J. D. West, "The evolution of large brain size in birds is related to social, not genetic, monogamy," *Biol J Linn Soc* 111, no. 3 (2014): 668–78.

124 **DNA analysis has revealed that extra-pair:** S. Griffith et al., "Extra pair paternity in birds: A review of interspecific variation and adaptive function," *Mol Ecol* 11 (2002): 2195–212.

125 **Take the Eurasian skylark:** J. Linossier et al., "Flight phases in the song of skylarks," *PLoS ONE* 8, no. 8 (2013): e72768.

125 **However, scientists found that 20 percent:** J. M. C. Hutchinson and S. C. Griffith, "Extra-pair paternity in the skylark, *Alauda arvensis,*" *Ibis* 150 (2008): 90–97.

125 **Behavioral ecologist Judy Stamps:** J. Stamps, "The role of females in extrapair copulations in socially monogamous territorial animals," in *Feminism and Evolutionary Biology: Boundaries, Intersections, and Frontiers*, ed. P. Gowaty (Washington, DC: Science, 1997), 294.

125 **A new theory offered by two biologists from the University of Norway:** S. Eliassen and C. Jørgensen, "Extra-pair mating and evolution of cooperative neighbourhoods," *PLoS ONE* 9, no. 7 (2014): e99878.

126 **(These findings echo earlier studies on western red-winged blackbirds):** E. M. Gray, "Female red-winged blackbirds accrue material benefits from copulating with extra-pair males," *Anim Behav* 53, no. 3 (1997): 625–39.

126 **"The reason that female birds engage":** Nancy Burley, personal communication, February 9, 2015.

127 **So even when he's guarding his mate:** J. Linossier et al., "Flight phases in the song of skylarks," *PLoS ONE* 8, no. 8 (2013): e72768. Researchers found that skylarks with shorter wings were more often cuckolded.

127 **Indeed, in species with more extra-pair paternity:** L. Z. Garamszegi et al., "Sperm competition and sexually size dimorphic brains in birds," *Proc R Soc B* 272 (2005): 159–66.

127 **According to one ornithologist, a favorite trick of the jay:** J. Mailliard, "California jays and cats," *Condor*, July 1904, 94–95.

127 **"The jay's ordinary alarm note":** L. D. Dawson, *The Birds of California: A Complete*

and Popular Account of the 580 Species and Subspecies of Birds Found in the State (San Diego: South Moulton Company, 1923).

128 **A scrub jay may lose up to 30 percent of his:** U. Grodzinski and N. S. Clayton, "Problems faced by food-caching corvids and the evolution of cognitive solutions," *Philos Trans R Soc Lond B* 365 (2010): 977–87.

128 **In a series of inspired studies, Nicola Clayton:** N. S. Clayton et al., "Social cognition by food-caching corvids: The western scrub-jay as a natural psychologist," *Philos Trans R Soc Lond B Biol Sci* 362, no. 1480 (2007): 507–22; J. M. Thom and N. S. Clayton, "Re-caching by western scrub-jays (*Aphelocoma californica*) cannot be attributed to stress," *PLoS ONE* 8, no. 1 (2013): e52936.

128 **If the observer can hear him:** G. Stulp et al., "Western scrub-jays conceal auditory information when competitors can hear but cannot see," *Biol Lett* 5 (2009): 583–85.

129 **In other words, say the researchers:** U. Grodzinski et al., "Peep to pilfer: What scrub-jays like to watch when observing others," *Anim Behav* 83 (2012): 1253–60.

129 **For Clayton and many others:** U. Grodzinski and N. S. Clayton, "Problems faced by food-caching corvids and the evolution of cognitive solutions," *Philos Trans R Soc Lond B* 365 (2010): 977–87.

129 **It's not clear whether caching:** Ibid.

130 **As Clayton and her colleague Nathan Emery warn:** N. J. Emery and N. S. Clayton, "Do birds have the capacity for fun?" *Curr Biol* 25, no. 1 (2015): R16–R19.

130 **They show off their social bonds with partners:** H. Fischer, "Das Triumphgeschrei der Graugans (*Anser anser*)," *Z Tierpsychol* 22 (1965): 247–304.

130 **A recent study at the Konrad Lorenz Research Station:** C. A. F. Wascher et al., "Heart rate during conflicts predicts post-conflict stress-related behavior in greylag geese," *PLoS ONE* 5, no. 12 (2010): e15751.

130 **These supremely social members of the crow family:** A. M. Seed et al., "Postconflict third-party affiliation in rooks, *Corvus frugilegus*," *Curr Biol* 17 (2007): 152–58.

131 **This was heralded by researchers:** N. J. Emery et al., "Cognitive adaptations to bonding in birds," *Philos Trans R Soc Lond B* 362 (2007): 489–505.

131 **Asian elephants were lately added to the list:** J. M. Plotnik and F. B. de Waal, "Asian elephants (*Elephas maximus*) reassure others in distress," *Peer J* 2 (2014): e278.

131 **Not long ago, Thomas Bugnyar:** O. Fraser and T. Bugnyar, "Do ravens show consolation? Responses to distressed others," *PLoS ONE* 5, no. 5 (2010): e10605.

132 **Then, for ten minutes after:** As a control in the experiment, the researchers observed the victims for ten minutes the day after a conflict to see whether other ravens approached them.

132 **These findings, they write:** O. Fraser and T. Bugnyar, "Do ravens show consolation? Responses to distressed others," *PLoS ONE* 5, no. 5 (2010): e10605.

132 **It was created by Teresa Iglesias:** T. Iglesias et al., "Western scrub-jay funerals: Cacophonous aggregations in response to dead conspecifics," *Anim Behav* 84, no. 5 (2012): 1103–11.

133 **Reactions to the study quickly cycled:** B. King, "Do birds hold funerals?" *13.7 Cosmos & Culture* (blog), NPR, September 6, 2012, http://www.npr.org/blogs/13.7/2012/09/06/160535236/do-birds-hold-funerals.

133 **In this sense, perhaps the scrub jay:** L. Erickson, "Scrub-jay funerals and blue jay Irish wakes," *Laura's Birding Blog*, September 26, 2012, http://webcache.googleusercontent.com/search?q=cache:http://lauraerickson.blogspot.com/2012/09/scrub-jay-funerals-and-blue-jay-irish.html.

133 **In a follow-up study, Iglesias and her colleagues:** T. L. Iglesias et al., "Dead heterospecifics as cues of risk in the environment: Does size affect response?" *Behaviour* 151 (2014): 1–22.

133 **This suggests that these gatherings:** Teresa Iglesias, personal communication, February 7, 2015.

134 **One definition of empathy:** M. L. Hoffman, "Is altruism part of human nature?" *J Personal Soc Psychol* 40 (1981): 121–37.

134 **Birds may not express emotions through the facial musculature:** N. J. Emery and N. S. Clayton, "Do birds have the capacity for fun?" *Curr Biol* 25, no. 1 (2015): R16–R19.

134 **Konrad Lorenz once noted that a greylag goose:** K. Lorenz, quoted in Marc Bekoff, "Grief in animals: It's arrogant to think we're the only animals who mourn" (blog), *Psychology Today*, October 29, 2009, http://www.psychologytoday.com/blog/animal-emotions/200910/grief-in-animals-its-arrogant-think-were-the-only-animals-who-mourn.

134 **Marc Bekoff, professor emeritus:** Ibid.

134 **In *Gifts of the Crow*, John Marzluff and Tony Angell:** *Gifts of the Crow* (New York: Free Press, 2013), 138–39.

135 **Marzluff has shown that when crows see:** D. J. Cross et al., "Distinct neural circuits underlie assessment of a diversity of natural dangers by American crows," *Proc R Soc B* 280 (2013): 20131046.

Five: Four Hundred Tongues: *Vocal Virtuosity*

137 **Although the president didn't dignify:** E. M. Halliday, *Understanding Thomas Jefferson* (New York: HarperCollins, 2001), 184. Apparently, writes Halliday, Jefferson was capable of both "childish delight" in a pet mockingbird and "icy ruthlessness" toward pet dogs owned by his slaves. At about the same time Jefferson called the mockingbird "a superior being," he heard from his Monticello overseer Edmund Bacon that dogs belonging to the slaves were killing some of his sheep and told him, "To secure wool enough, the negroes' dogs must all be killed. Do not spare a single one."

137 **"I sincerely congratulate you":** Writing from Monticello in May 1793, Thomas Mann Randolph informed Jefferson in Philadelphia of the advent of the first resident mock-

ingbird, and Jefferson responded with his well-known tribute to *Mimus polyglottos*, http://www.monticello.org/site/research-and-collections/mockingbirds#_note-1.

138 **unlovely avian expletive that one naturalist:** J. Lembke, *Dangerous Birds* (New York: Lyons & Burford, 1992), 66.

138 **He knew he could mimic other birds:** T. Jefferson in a letter to Abigail Adams, June 21, 1785.

138 **Inside the Lohrfink Auditorium:** Society for Neuroscience conference on "Birdsong: Rhythms and clues from neurons to behavior," November 14–15, 2014, Georgetown University, Washington DC (hereafter SFN conference).

138 **It's called vocal learning:** C. I. Petkov et al., "Birds, primates, and spoken language origins: Behavioral phenotypes and neurobiological substrates," *Front Evol Neurosci* 4 (2012): 12; E. D. Jarvis, "Evolution of brain pathways for vocal learning in birds and humans," in *Birdsong, Speech, and Language*, ed. J. J. Bolhuis and M. Everaert (Cambridge, MA: MIT Press, 2013), 63–107; D. Kroodsma et al., "Behavioral evidence for song learning in the suboscine bellbirds (*Procnias* spp.; Cotingidae)," *Wilson J Ornithol* 125, no. 1 (2013): 1–14.

138 **If cognition is defined:** S. J. Shettleworth, *Cognition, Evolution, and Behavior* (New York: Oxford University Press, 2010), 23.

139 **The scientists are noting the remarkable:** A. R. Pfenning et al., "Convergent transcriptional specializations in the brains of humans and song-learning birds," *Science* 346, no. 6215 (2014): 13333.

139 **(they stutter, for instance):** L. Kubikova et al., "Basal ganglia function, stuttering, sequencing, and repair in adult songbirds," *Sci Rep* 13, no. 4 (2014): 6590.

139 **Johan Bolhuis, a neurobiologist:** J. Bolhuis, "Birdsong, speech and language" (SFN conference presentation, November 14–15, 2014).

140 **On the voyage of the *Beagle*:** C. Darwin, *Voyage of the Beagle*, 1839 (New York: Penguin Classics, 1989).

141 **"It's not unlike our singing":** L. Riters, "Why birds sing: The neural regulation of the motivation to communicate" (SFN conference presentation, November 14–15, 2014).

141 **"By studying vocal learning":** Quotes from Erich Jarvis from interview with Jarvis, March 23, 2012; E. Jarvis, "Identifying analogous vocal communication regions between songbird and human brains" (SFN conference presentation, November 14–15, 2014).

142 **In the open, sound travels best:** E. Nemeth et al., "Differential degradation of antbird songs in a neotropical rainforest: Adaptation to perch height?" *Jour Acoust Soc Am* 110 (2001): 3263–74.

142 **Those singing on the forest floor:** H. Slabbekoorn, "Singing in the wild: The ecology of birdsong," in *Nature's Music: The Science of Birdsong*, ed. P. Marler and H. Slabbekoorn (Amsterdam: Elsevier Academic Press, 2004).

142 **Some use frequencies that avoid:** M. J. Ryan et al., "Cognitive mate choice," in *Cog-*

nitive Ecology II, ed. R. Dukas and J. Ratcliffe (Chicago: University of Chicago Press, 2009), 137–55.

142 **Birds living near airports sing:** D. Gil et al., "Birds living near airports advance their dawn chorus and reduce overlap with aircraft noise," *Behav Ecol* 26, no. 2 (2014): 435–43.

142 **Scientists took a long time:** R. A. Suthers and S. A. Zollinger, "Producing song: The vocal apparatus," in *Behavioral Neurobiology of Bird Song*, ed. H. P. Zeigler and P. Marler (New York: Annals of the New York Academy of Sciences, 2014), 109–29.

142 **Only in the past few years:** D. N. Düring et al., "The songbird syrinx morphome: A three-dimensional, high-resolution, interactive morphological map of the zebra finch vocal organ," *BMC Biol* 11 (2013): 1.

143 **Gifted songbirds such as the mockingbird:** S. A. Zollinger et al., "Two-voice complexity from a single side of the syrinx in northern mockingbird *Mimus polyglottos* vocalizations," *J Exp Biol* 211 (2008): 1978–91.

143 **Certain songbirds, such as:** C. P. H. Elemans et al., "Superfast vocal muscles control song production in songbirds," *PLoS ONE* 3, no. 7 (2008): e2581.

143 **The winter wren, a little:** http://bna.birds.cornell.edu/bna/species/720doi:10.2173.

143 **Birds with a more elaborate:** However, parrots and lyrebirds, both famous for their vocal versatility, seem to make do with just a few.

144 **Not such an easy task:** T. Gentner, "Mechanisms of auditory attention" (SFN conference presentation, November 14–15, 2014).

144 **Sonograms comparing a prototype:** D. Kroodsma, *The Singing Life of Birds* (Boston: Houghton Mifflin, 2007), 76–77.

144 **Scientists found that when a mockingbird:** S. A. Zollinger and R. A. Suthers, "Motor mechanisms of a vocal mimic: Implications for birdsong production," *Proc R Soc B* 271 (2004): 483–91.

145 **And if he's facing a too-rapid-fire delivery:** L. A. Kelley et al., "Vocal mimicry in songbirds," *Anim Behav* 76 (2008): 521–28.

145 **A cousin *Mimidae*, the brown thrasher:** D. E. Kroodsma and L. D. Parker, "Vocal virtuosity in the brown thrasher," *Auk* 94 (1977): 783–85.

145 **Common European starlings:** H. Hultsch and D. Todt, "Memorization and reproduction of songs in nightingales (*Luscinia megarhynchos*): Evidence for package formation," *J Comp Phys A* 165 (1989): 197–203.

145 **Marsh warblers are known:** F. Dowsett-Lemaire, "The imitative range of the song of the marsh warbler *Acrocepalus palustris*, with special reference to imitations of African birds," *Ibis* 121 (2008): 453–68.

145 **As one naturalist noted:** H. J. Pollock, "Living with the lyrebirds," *Proc Zool Soc* (July 23, 1965): 20–24.

145 **The fork-tailed drongo:** T. P. Flower, "Deceptive vocal mimicry by drongos," *Proc R Soc B* (2010), doi:10.1098/rspb.2010.1932.

145 **There are reports of a bullfinch:** P. Marler and H. Slabbekoorn, *Nature's Music: The Science of Birdsong* (Amsterdam: Elsevier Academic Press, 2004), 35.

146 **The *New Yorker* once reported:** W. C. Fitzgibbon, "Talk of the Town," *New Yorker*, August 14, 1954.

146 **Parrots are unusual:** V. R. Ohms et al., "Vocal tract articulation revisited: the case of the monk parakeet," *J Exp Biol* 215 (2012): 85–92; G. J. L. Beckers et al., "Vocal-tract filtering by lingual articulation in a parrot," *Curr Biol* 14, no. 7 (2004): 1592–97.

146 **Irene Pepperberg made African greys:** I. M. Pepperberg, *The Alex Studies* (Cambridge, MA: Harvard University Press, 1999), 13–52.

146 **He also cottoned to phrases:** Irene Pepperberg, personal communication, May 8, 2015.

147 **Not long ago, a naturalist:** Naturalist Martyn Robinson's story was reported in H. Price, "Birds of a feather talk together," *Aust Geogr*, September 15, 2011, Iangeo graphic.com.au/news/2011/09/birds-of-a-feather-talk-together/.

147 **A tally of one mockingbird's tunes:** D. Kroodsma, *The Singing Life of Birds*, 70.

147 **The Arnold Arboretum mocker:** C. H. Early, "The mockingbird of the Arnold Arboretum," *Auk* 38 (1921): 179–81.

148 **So particular is a song:** R. D. Howard, "The influence of sexual selection and interspecific competition on mockingbird song," *Evolution* 28, no. 3 (1974): 428–38; J. L. Wildenthal, "Structure in primary song of mockingbird," *Auk* 82 (1965): 161–89; J. J. Hatch, "Diversity of the song of mockingbirds reared in different auditory environments" (PhD thesis, Duke University, 1967).

148 **Mockingbirds regularly imitate:** K. C. Derrickson, "Yearly and situational changes in the estimate of repertoire size in northern mockingbirds (*Mimus polyglottos*)," *Auk* 104 (1987): 198–207.

148 **The fancifully named "Beau Geste":** J. R. Krebs, "The significance of song repertoires: The Beau Geste hypothesis," *Anim Behav* 25, no. 2 (1977): 475–78.

149 **As ornithologist J. Paul Visscher:** J. P. Visscher, "Notes on the nesting habits and songs of the mockingbird," *Wilson Bulletin* 40 (1928): 209–16.

149 **To sort out the nature-nurture:** A. Laskey, "A mockingbird acquires his song repertory," *Auk* 61 (1944): 211–19.

149 **(His observer was one of those scientists):** http://naturalhistorynetwork.org/journal/articles/8-donald-culross-peatties-an-almanac-for-moderns/.

150 **The ideal model organism:** E. Kandel quoting fruit fly behavior specialist Chip Quinn in *In Search of Memory* (New York: W. W. Norton, 2006), 148.

150 **The zebra finch does:** R. Zann, *The Zebra Finch: A Synthesis of Field and Laboratory Studies* (New York: Oxford University Press, 1996).

151 **"Because it's impractical":** R. Mooney, "Translating birdsong research" (SFN conference presentation, November 14–15, 2014).

151 **A baby zebra finch:** Discussion of the process of song learning in birds comes from

S. Nowicki and W. A. Searcy, "Song function and the evolution of female preferences: Why birds sing and why brains matter," *Ann N Y Acad Sci* 1016 (June 2004): 704–23.

151 **Incidentally, birds do have ears:** R. Dooling, "Audition: Can birds hear everything they sing?" in *Nature's Music: The Science of Birdsong*, ed. P. Marler and H. Slabbekoorn (Amsterdam: Elsevier Academic Press, 2004), 206–25.

151 **(If a bird's hair cells):** J. S. Stone and D. A. Cotanche, "Hair cell regeneration in the avian auditory epithelium," *Int J Deve Biol* 51, no. 607 (2007): 633–47.

152 **In one region, the high vocal:** J. F. Prather et al., "Neural correlates of categorical perception in learned vocal communication," *Nat Neurosci* 12, no. 2 (2009): 221–28.

152 **By the time the young bird:** P. Ardet et al., "Song tutoring in pre-singing zebra finch juveniles biases a small population of higher-order song selective neurons towards the tutor song," *J Neurophysiol* 108, no. 7 (2012): 1977–87.

152 **It's a perfect example of the intertwining:** J. J. Bolhuis et al., "Twitter evolution: Converging mechanisms in birdsong and human speech," *Nat Rev Neurosci* 11 (2010): 747–59.

152 **This discovery—that some young birds:** Ibid.

153 **One neuroscientist, Sarah London:** S. London, "Mechanisms for sensory song learning" (SFN conference presentation, November 14–15, 2014).

153 **In the first two or three years of life:** P. K. Kuhl, "Learning and representation in speech and language," *Curr Opin Neurobiol* 4, no. 6 (1994): 812–22.

153 **After puberty, we have to work:** J. J. Bolhuis et al., "Twitter evolution: Converging mechanisms in birdsong and human speech," *Nat Rev Neurosci* 11 (2010): 747–48.

154 **Scientists have discovered:** D. Aronov et al., "A specialized forebrain circuit for vocal babbling in the juvenile songbird," *Science* 320 (2008): 630–34.

154 **Dopamine may provide:** K. Simonyan et al., "Dopamine regulation of human speech and bird song: A critical review," *Brain Lang* 122, no. 3 (2012): 142–50.

155 **Sleep seems to play a role:** S. Derégnaucourt et al., "How sleep affects the developmental learning of bird song," *Nature* 433 (2005): 710–16; S. S. Shank and D. Margoliash, "Sleep and sensorimotor integration during early vocal learning in a songbird," *Nature* 458 (2009): 73–77.

155 **Who is listening makes:** S. C. Woolley and A. Doupe, "Social context-induced song variation affects female behavior and gene expression," *PLoS Biol* 6, no. 3 (2008): e62.

155 **"I've listened to the two versions":** R. Mooney, "Translating birdsong research" (SFN conference presentation, November 14–15, 2014).

155 **Brain-imaging studies by Erich Jarvis:** E. D. Jarvis et al., "For whom the bird sings: Context-dependent gene expression," *Neuron* 21 (1998): 775–88.

156 **Mother finches also guide:** http://babylab.psych.cornell.edu/wp-content/uploads/2012/12/newsletter_fall_2012.pdf.

156 **All of this is powerful proof:** M. H. Goldstein, "Social interaction shapes babbling: Testing parallels between birdsong and speech," *PNAS* 100, no. 13 (2003): 8030–35.

156 **But then, Fernando Nottebohm:** F. Nottebohm, "The neural basis of birdsong," *PLoS Biol* 3, no. 5 (2005): e164.

156 **Not only do birds:** A. J. Doupe and P. K. Kuhl, "Birdsong and human speech: Common themes and mechanisms," *Annu Rev Neurosci* 22 (1999): 567–631; J. J. Bolhuis et al., "Twitter evolution: Converging mechanisms in birdsong and human speech," *Nat Rev Neurosci* 11 (2010): 747–48; P. Marler, "A comparative approach to vocal learning: Song development in white-crowned sparrows," *J Comp Physiol Psych* 7, no. 2, pt. 2 (1970): 1–25; F. Nottebohm, "The origins of vocal learning," *Amer Natur* 106 (1972): 116–40.

157 **A new theory by Shigeru Miyagawa:** S. Miyagawa et al., "The integration hypothesis of human language evolution and the nature of contemporary languages," *Front Psychol* 5 (2014): 564.

157 **In Miyagawa's view, human language:** S. Miyagawa et al., "The emergence of hierarchical structure in human language," *Front Psychol* 4 (2013): 71.

157 **But what's really similar:** Erich Jarvis, interview, March 23, 2012.

158 **And indeed, that afternoon in Georgetown:** The team found that this similar gene expression was most pronounced in two parallel parts of the songbird brain and the human brain: in Area X of the songbird brain, a "striatal" region necessary for vocal learning, and in the human striatum activated during speech production; as well as in a part of the bird brain called the RA (robust nucleus of the arcopallium) analog, necessary for song production, and in the laryngeal motor cortex regions in humans that control speech production. See A. R. Pfenning et al., "Convergent transcriptional specializations in the brains of humans and song-learning birds," *Science* 346, no. 6215 (2014): 13333.

158 **In one recent imaging study:** Interview with Erich Jarvis; G. Feenders et al., "Molecular mapping of movement-associated areas in the avian brain: A motor theory for vocal learning origin," *PLoS ONE* 3, no. 3 (2008): e1768.

159 **"It's a case of convergence":** J. Bolhuis, "Birdsong, speech and language" (SFN conference presentation, November 14–15, 2014).

159 **In this way, vocal learning:** G. Zhang et al., "Comparative genomics reveals insights into avian genome evolution and adaptation," *Science* 346, no. 6215 (2014): 1311–19.

159 **Interesting to note:** Recent DNA analysis suggests that parrots may be more closely related to songbirds than once thought. See S. J. Hackett et al., "A phylogenomic study of birds reveals their evolutionary history," *Science* 320, no. 5884 (2008): 1763–68; E. D. Jarvis et al., "Whole genome analyses resolve early branches in the tree of life of modern birds," *Science* 346, no. 6215 (2014): 1320–31; H. Horita et al., "Specialized motor-driven dusp1 expression in the song systems of multiple lineages of vocal learning birds," *PLoS ONE* 7, no. 8 (2012): e42173. "These findings led to the novel proposal that vocal learning evolved twice in birds (once in hummingbirds and again in the common ancestor of songbirds and parrots) and was subsequently lost in suboscine songbirds," write the scientists.

159 **Parrots have a kind:** M. Chakraborty et al., "Core and shell song systems unique to the parrot brain," *PLoS ONE* (in press, 2015).

160 **This, says Jarvis, may be:** Erich Jarvis, interview; E. D. Jarvis, "Selection for and against vocal learning in birds and mammals," *Ornith Sci* 5 (special issue on the neuroecology of birdsong, 2006): 5–14.

160 **Jarvis suspects that vocal learning:** G. Arriago and E. D. Jarvis, "Mouse vocal communication system: Are ultrasounds learned or innate?" *Brain Lang* 124 (2013): 96–116.

160 **Research by Kazuo Okanoya:** Erich Jarvis, interview; H. Kagawa et al., "Domestication changes innate constraints for birdsong learning," *Behav Proc* 106 (2014): 91–97; K. Okanoya, "The Bengalese finch: A window on the behavioral neurobiology of birdsong syntax study," *Ann N Y Acad Sci* 1016 (2006): 724–35; K. Suzuki et al., "Behavioral and neural trade-offs between song complexity and stress reaction in a wild and domesticated finch strain," *Neurosci Biobehav Rev* 46, pt. 4 (2014): 547–56.

161 **Because songs well sung:** Erich Jarvis interview; see also L. Z. Garamszegi et al., "Sexually size dimorphic brains and song complexity in passerine birds," *Behav Ecol* 16, no. 2 (2004): 335–45.

162 **For a long time scientists:** There is some evidence for this. A study of song sparrows on a rocky island in British Columbia found that males with larger repertoires were more likely to mate during their first year and that females that mated with males with larger repertoires bred earlier. J. M. Reid et al., "Song repertoire size predicts initial mating success in male song sparrows, *Melospiza melodia*," *Anim Behav* 68, no. 5 (2004): 1055–63.

162 **Studies show that females:** J. Podos, "Sexual selection and the evolution of vocal mating signals: Lessons from neotropical birds," in *Sexual Selection: Perspectives and Models from the Neotropics*, ed. R. H. Macedo and G. Machado (Amsterdam: Elsevier Academic Press, 2013), 341–63.

162 **Many songbirds have regional:** J. Podos and P. S. Warren, "The evolution of geographic variation in birdsong," *Adv Stud Behav* 37 (2007): 403–58. For the first few weeks of life, a young sparrow can learn a new dialect. But after he reaches three months or so, training fails to have any effect. His song is set.

163 **According to ornithologist Donald Kroodsma:** J. Uscher, "The Language of Song: An Interview with Donald Kroodsma," *Scientific American*, July 1, 2002, https://www.scientificamerican.com/articl/the-language-of-song-an-I.

163 **The geographical separation:** P. Marler and M. Tamura, "Song 'dialects' in three populations of white-crowned sparrows," *Condor* 64 (1962): 368–77.

163 **Some time ago, Robert Payne:** R. B. Payne et al., "Biological and cultural success of song memes in indigo buntings," *Ecology* 69 (1988): 104–17.

163 **And here's the point:** J. M. Lapierre, "Spatial and age-related variation in use of locally common song elements in dawn singing of song sparrows *Melospiza melodia*: Old males sing the hits," *Behav Ecol Sociobiol* 65 (2011): 2149–60.

164 **Richard Mooney makes the case:** R. Mooney, "Translating birdsong research."

164 **Lab studies show that zebra finch:** S. C. Woolley and A. J. Doupe, "Social context-induced song variation affects female behavior and gene expression," *PLoS Biol* 6 (2008): e62.

164 **Male great reed warblers:** E. Wegrzyn et al., "Whistle duration and consistency reflect philopatry and harem size in great reed warblers," *Anim Behav* 79 (2010): 1363–92.

165 **Likewise, male banded wrens:** E. R. A. Cramer et al., "Infrequent extra-pair paternity in banded wrens," *Condor* 112 (2011): 637–45; B. E. Byers, "Extrapair paternity in chestnut-sided warblers is correlated with consistent vocal performance," *Behav Ecol* 18 (2007): 130–36.

165 **The same holds true:** C. A. Botero et al., "Syllable type consistency is related to age, social status, and reproductive success in the tropical mockingbird," *Anim Behav* 77, no. 3 (2009): 701–6.

165 **Scientists are still sorting out:** The following discussion of song signaling draws on personal communication with Neeltje Boogert, April 2015.

165 **Sexy syllables in canaries:** R. A. Suthers et al., "Bilateral coordination and the motor basis of female preference for sexual signals in canary song," *J Exp Biol* 215 (2015): 2950–59.

165 **Listening for supersexy:** Ibid.

165 **It goes back to those critical:** S. Nowicki and W. A. Searcy, "Song function and the evolution of female preferences: Why birds sing, why brains matter," *Ann N Y Acad Sci* 1016 (2004): 704–23.

165 **If something happens during:** S. Nowicki et al., "Brain development, song learning and mate choice in birds: A review and experimental test of the 'nutritional stress hypothesis,'" *J Comp Physiol A* 188 (2002): 1003–14; S. Nowicki et al., "Quality of song learning affects female response to male bird song," *Proc R Soc B* 269 (2002): 1949–54.

166 **One study, for example:** H. Brumm et al., "Developmental stress affects song learning but not song complexity and vocal amplitude in zebra finches," *Behav Ecol Sociobiol* 63, no. 9 (2009): 1387–95.

166 **This "cognitive capacity hypothesis":** N. J. Boogert et al., "Song complexity correlates with learning ability in zebra finch males," *Anim Behav* 76 (2008): 1735–41; C. N. Templeton et al., "Does song complexity correlate with problem-solving performance in flocks of zebra finches?" *Anim Behav* 92 (2014): 63–71.

166 **When Neeltje Boogert:** N. J. Boogert et al., "Song complexity correlates with learning ability in zebra finch males," *Anim Behav* 76 (2008): 1735–41; N. J. Boogert et al., "Mate choice for cognitive traits: A review of the evidence in nonhuman vertebrates," *Behav Ecol* 22 (2011): 447–59.

166 **When Boogert and her colleagues:** N. J. Boogert et al., "Song repertoire size in male song sparrows correlates with detour reaching, but not with other cognitive measures," *Anim Behav* 81 (2011): 1209–16.

166 **And recently, in a study of zebra finches:** C. N. Templeton et al., "Does song complexity correlate with problem-solving performance in flocks of zebra finches?" *Anim Behav* 92 (2014): 63–71.

167 **Confounding factors may muddy:** Neeltje Boogert, personal communication, April 2015.

167 **Not long ago, Carlos Botero:** C. A. Botero et al., "Climatic patterns predict the elaboration of song displays in mockingbirds," *Curr Biol* 19, no. 13 (2009): 1151–55.

167 **In fickle environments:** C. A. Botero and S. R. de Kort, "Learned signals and consistency of delivery: A case against receiver manipulation in animal communication," in *Animal Communication Theory: Information and Influence*, ed. U. Stegmann (New York: Cambridge University Press, 2013), 281–96; C. A. Botero et al., "Syllable type consistency is related to age, social status and reproductive success in the tropical mockingbird," *Anim Behav* 77, no. 3 (2009): 701–6.

168 **As ornithologist Donald Kroodsma explains:** D. Kroodsma, *The Singing Life of Birds*, 201; Donald Kroodsma, interview with *Birding*, www.aba.org/birding/v4ln3p18wl.pdf.

168 **It's known as the mating-mind hypothesis:** G. F. Miller, *The Mating Mind: How Sexual Choice Shaped the Evolution of Human Nature* (New York: Doubleday, 2000); T. W. Fawcett et al., "Female assessment: Cheap tricks or costly calculations," *Behav Ecol* 22, no. 3 (2011): 462–63.

168 **Birds that sing their songs:** T. D. Sasaki et al., "Social context-dependent singing-regulated dopamine," *J Neurosci* 26 (2006): 9010–14.

168 **To find out which season's song:** L. Riters, "Why birds sing: The neural regulation of the motivation to communicate" (SFN conference presentation, November 14–15, 2014).

Six: The Bird Artist: *Aesthetic Aptitude*

171 **On a sun-mottled spot:** Discussion of bowerbird behavior and displays draws on the research of Gerald Borgia and Jason Keagy; my interview with Gerald Borgia on July 6, 2012, and personal communication with Borgia on February 13, 2015; personal communication with Jason Keagy, March 16, 2015; G. Borgia, "Why do bowerbirds build bowers?" *American Scientist* 83 (1995): 542e547.

172 **Watch for a few more days:** R. E. Hicks et al., "Bower paint removal leads to reduced female visits, suggesting bower paint functions as a chemical signal," *Anim Behav* 85 (2013): 1209–15.

172 **especially the ornate constructions:** P. Goodfellow, *Avian Architecture* (Princeton, NJ: Princeton University Press, 2011), 102.

173 **"The implement which determines":** J. Michelet, *The Birds*, 1869, 248–50, www.gutenberg.org/eboks/43341.

173 **The nest was anchored:** New Zealand Birds: http://www.nzbirds.com/birds/fantailnest.html#sthash.

173 **Its nest is a flexible bag:** M. Hansell, *Animal Architecture* (Oxford: Oxford University Press, 2005), 36, 71.

173 **"A bird's nest is the most graphic":** C. Dixon, *Birds' Nests: An Introduction to the Science of Caliology* (London: Grant Richards, 1902), v.

173 **Nobel laureate Niko Tinbergen:** W. H. Thorpe, *Learning and Instinct in Animals* (London: Methuen, 1956), 36.

174 **but then remarked how amazed:** M. Hansell, *Animal Architecture*, 71.

174 **The magnificent creation:** A. McGowan et al., "The structure and function of nests of long-tailed tits *Aegithalos caudatus*," *Func Ecol* 18, no. 4 (2004): 578–83.

174 **It makes sense, then:** Z. J. Hall et al., "Neural correlates of nesting behavior in zebra finches (*Taeniopygia guttata*)," *Behav Brain Res* 264 (2014): 26–33.

174 **In an experiment reported in 2014:** I. E. Bailey et al., "Physical cognition: Birds learn the structural efficacy of nest material," *Proc R Soc B* 281, no. 1784 (2014): 20133225.

174 **In the wild, zebra finches:** R. Zann, *The Zebra Finch: A Synthesis of Field and Laboratory Studies* (New York: Oxford University Press, 1996).

174 **To see whether the birds:** I. E. Bailey et al., "Birds build camouflaged nests," *Auk* 132 (2015): 11–15.

174 **Village weaverbirds, too, learn:** E. C. Collias and N. E. Collias, "The development of nest-building behavior in a weaverbird," *Auk* 81 (1964): 42–52.

175 **So remarkable is the bowerbird:** E. T. Gilliard, *Birds of Paradise and Bower Birds* (Boston: D. R. Godine, 1979).

175 **As soon as she lands:** Description of bowerbird dance and vocal display draws on the research of Gerald Borgia and Jason Keagy; my interview with Gerald Borgia on July 6, 2012, and personal communication with Borgia on February 13, 2015; personal communication with Jason Keagy, March 16, 2015.

177 **These bowerbirds exhibit:** Gerald Borgia, interview, July 6, 2012.

177 **Indeed, in every aspect:** Ibid.

177 **Sometimes they prune:** A. F. Larned et al., "Male satin bowerbirds use sunlight to illuminate decorations to enhance mating success," Front Behav Neurosci conference abstract: Tenth International Congress of Neuroethology (2012), doi:10.3389/conf.fnbeh.2012.27.00372.

177 **"In templating, a male":** Gerald Borgia, interview; J. Keagy et al., "Cognitive ability and the evolution of multiple behavioral display traits," *Behav Ecol* 23 (2011): 448–56.

178 **When experimenters messed with the bowers:** J. Keagy et al., "Complex relationship between multiple measures of cognitive ability and male mating success in satin bowerbirds, *Ptilonorhynchus violaceus*," *Anim Behav* 81 (2011): 1063–70.

178 **The Vogelkop bowerbird builds:** P. Rowland, *Bowerbirds* (Melbourne: CSIRO Publishing, 2008).

179 **The reddish light from the sticks:** J. A. Endler et al., "Visual effects in great bowerbird sexual displays and their implications for signal design," *Proc R Soc B* 281 (2014): 20140235.

179 **According to John Endler:** John Endler, personal communicaton, January 18 and February 3, 2015; J. A. Endler et al., "Great bowerbirds create theaters with forced perspective when seen by their audience," *Curr Biol* 20, no. 18 (2010): 1679–84.

180 **It could be a simple matter:** John Endler, personal communicaton, January 18 and February 3, 2015.

180 **One thing we know:** Endler quoted at http://www.deakin.edu.au/research/stories /2012/01/23/males-up-to-their-old-tricks.

180 **The birds are deeply committed:** L. A. Kelley and J. A. Endler, "Male great bowerbirds create forced perspective illusions with consistently different individual quality," *PNAS* 109, no. 51 (2012): 20980–85.

181 **Surveys suggest that blue:** S. E. Palmer and K. B. Schloss, "An ecological valence theory of human color preference," *PNAS* 107, no. 19 (2010): 8877–82.

181 **In nature blue is unusual:** J. T. Bagnara et al., "On the blue coloration of vertebrates," *Pigment Cell Res* 20, no. 1 (2007): 14–26.

181 **Borgia's research team uses:** See "Destruction and stealing" video at http://www .life.umd.edu/biology/borgialab/#Videos.

182 **Some observers even go:** A. J. Marshall, "Bower-birds," *Biol Rev* 29, no. 1 (1954): 1–45.

182 **The satin bowerbirds' urge:** J. Keagy et al., "Male satin bowerbird problem-solving ability predicts mating success," *Anim Behav* 78 (2009): 809–17; J. Keagy et al., "Complex relationship between multiple measures of cognitive ability and male mating success in satin bowerbirds, *Ptilonorhynchus violaceus*," *Anim Behav* 81 (2011): 1063–70; J. Keagy et al., "Cognitive ability and the evolution of multiple behavioral display traits," *Behav Ecol* 23 (2012): 448–56.

182 **Most of the birds that solved the puzzle:** See Jason Keagy video, https://www .youtube.com/watch?v=kn0VsIdD1AA.

183 **John Endler suggests:** J. Endler, "Bowerbirds, art and aesthetics," *Commun Integr Biol* 5, no. 3 (2012): 281–83.

183 **Richard Prum, an ornithologist:** R. O. Prum, "Coevolutionary aesthetics in human and biotic artworlds," *Biol Phil* 28, no. 5 (2014): 811–32.

183 **The naturalist and filmmaker Heinz Sielmann:** K. von Frisch, *Animal Architecture* (New York: Harcourt Brace, 1974), 243–44.

184 **According to Gerald Borgia:** G. Borgia and J. Keagy, "Cognitively driven co-option and the evolution of complex sexual display in bowerbirds," in *Animal Signaling and Function: An Integrative Approach*, ed. D. Irschick et al. (New York: John Wiley and Sons, 2015), 75–101; Jason Keagy, personal communication, March 16, 2015.

184 **Gail Patricelli, an animal behaviorist:** Gail Patricelli, personal communication, March 8, 2015.

184 **To see how different males handle:** G. L. Patricelli et al., "Male satin bowerbirds, *Ptilonorhynchus violaceus*, adjust their display intensity in response to female startling: An experiment with robotic females," *Anim Behav* 71 (2006): 49–59; G. Patricelli et al., "Male displays adjusted to female's response: Macho courtship by the satin bowerbird is tempered to avoid frightening the female," *Nature* 415 (2002): 279–80.

185 **And, as with song learning:** S. Nowicki et al., "Brain development, song learning and mate choice in birds: A review and experimental test of the 'nutritional stress hypothesis,'" *J Comp Physiol A* 188 (2002): 1003–14; S. Nowicki et al., "Quality of song learning affects female response to male bird song," *Proc R Soc B* 269 (2002): 1949–54.

185 **"Young males build":** Gerald Borgia, interview, July 6, 2012.

186 **"Also, the juveniles":** Jason Keagy, personal communication, March 16, 2015.

186 **(When experimenters removed this paint):** R. E. Hicks, "Bower paint removal leads to reduced female visits, suggesting bower paint functions as a chemical signal," *Anim Behav* 85 (2013): 1209–15. .

186 **A choosy female, for her part:** J. Keagy et al., "Male satin bowerbird problem-solving ability predicts mating success," *Anim Behav* 78 (2009): 809–17; J. Keagy et al., "Complex relationship between multiple measures of cognitive ability and male mating success in satin bowerbirds, *Ptilonorhynchus violaceus*," *Anim Behav* 81 (2011): 1063–70.

186 **As Jason Keagy observes:** Jason Keagy, personal communication, March 16, 2015; C. Rowe and S. D. Healy, "Measuring variation in cognition," *Behav Ecol* (2014), doi:10.1093/beheco/aru090.

187 **Then she must compare:** Female bowerbirds recall information about mates from previous years. See J. A. C. Uy et al., "Dynamic mate-searching tactic allows female satin bowerbirds *Ptilonorhynchus violaceus* to reduce searching," *Proc R Soc B* 267 (2000): 251–56.

187 **"It turns out to be very similar":** Gail Patricelli, personal communication, March 8, 2015.

187 **The many display traits:** J. Keagy et al., "Cognitive ability and the evolution of multiple behavioral display traits," *Behav Ecol* 23 (2012): 448–56; G. Borgia, "Bower quality, number of decorations and mating success of male satin bowerbirds (*Ptilonorhynchus violaceus*): An experimental analysis," *Anim Behav* 33 (1985): 266–71; C. A. Loffredo and G. Borgia, "Male courtship vocalizations as cues for mate choice in the satin bowerbird (*Ptilonorhynchus violaceus*)," *Auk* 103 (1986): 189–95.

188 **(As it happens, research suggests):** M. D. Prokosch, "Intelligence and mate choice: Intelligent men are always appealing," *Evol Hum Behav* 30 (2009): 11–20.

188 **This was Charles Darwin's:** R. O. Prum, "Aesthetic evolution by mate choice: Darwin's *really* dangerous idea," *Philos Trans R Soc Lond B* 367 (2012): 2253–65.

188 **As Ronald Fisher suggested:** This is the so-called runaway sexual selection model, or

the "sexy son" model, because the main benefit females gain from their choice is sexier sons who mate more often, passing on the genes for sexy traits in males and preferences for these traits in females. Gail Patricelli, personal communication, March 8, 2015.

188 **Males may gradually evolve:** C. Darwin, *The Descent of Man* (London: John Murray, 1871), 793.

189 **Some years ago, Watanabe:** S. Watanabe, "Animal aesthetics from the perspective of comparative cognition," in S. Watanabe and S. Kuczaj, eds, *Emotions of Animals and Humans* (Tokyo: Springer, 2012), 129; S. Watanabe et al., "Discrimination of paintings by Monet and Picasso in pigeons," *J Exp Anal Behav* 63 (1995): 165–74; S. Watanabe, "Van Gogh, Chagall and pigeons," *Anim Cogn* 4 (2001): 147–51.

189 **To probe whether birds:** S. Watanabe, "Pigeons can discriminate 'good' and 'bad' paintings by children," *Anim Cogn* 13, no. 1 (2010): 75–85.

190 **To find out, Watanabe's team:** Y. Ikkatai and S. Watanabe, "Discriminative and reinforcing properties of paintings in Java sparrows (*Padda oryzivora*)," *Anim Cogn* 14, no. 2 (2011): 227–34.

190 **But Watanabe's work:** S. Watanabe, "Discrimination of painting style and beauty: Pigeons use different strategies for different tasks," *Anim Cogn* 14, no. 6 (2011): 797–808.

190 **Pigeons shown a series:** R. E. Lubow, "High-order concept formation in the pigeon," *J Exp Anal Behav* 21 (1973): 475–83.

190 **They can also recognize:** C. Stephan et al., "Have we met before? Pigeons recognize familiar human face," *Avian Biol Res* 5, no. 2 (2012): 75.

191 **In the hopes of finding:** J. Barske et al., "Female choice for male motor skills," *Proc R Soc B* 278, no. 1724 (2011): 3523–28.

191 **When scientists looked:** L. B. Day et al., "Sexually dimorphic neural phenotypes in golden-collared manakins," *Brain Behav Evol* 77 (2011): 206–18.

191 **Further research into several:** W. R. Lindsay et al., "Acrobatic courtship display coevolves with brain size in manakins (*Pipridae*)," *Brain Behav Evol* (2015), doi: 10.1159/000369244.

192 **Three species, however:** Gerald Borgia, personal communication; B. J. Coyle et al., "Limited variation in visual sensitivity among bowerbird species suggests that there is no link between spectral tuning and variation in display colouration," *J Exp Biol* 215 (2012): 1090–1105.

192 **Still, some of the cues birds:** For instance, animals of all kinds favor partners with a balance, a mirror image, between the two sides of the body. This makes good sense. Symmetry in nature nearly always signals important information. In plants and animals, it's often a sign of health, as it suggests freedom from the mutations, diseases, and environmental stresses that undo health, such as extreme temperatures or food scarcity.

192 **Experiments in the 1950s:** B. Rensch, "Die wirksamkeit ästhetischer faktoren bei wirbeltieren," *Z Tierpsychol* 15 (1958): 447–61.

192 **Nobel laureate Karl von Frisch once wrote:** K. von Frisch, *Animal Architecture* (New York: Harcourt Brace, 1974), 244.

Seven: A Mapping Mind: *Spatial (and Temporal) Ingenuity*

195 **That's pretty much what happened:** K. Thorup et al., "Evidence for a navigational map stretching across the continental U.S. in a migratory songbird," *PNAS* 104, no. 46 (2008): 18115–19.

196 **As Julia Frankenstein:** J. Frankenstein, "Is GPS All in Our Heads?" *New York Times*, Sunday Review, February 2, 2012.

197 **Sometimes known as:** Information on pigeon racing comes from W. M. Levi, *The Pigeon* (Sumter, SC: Levi Publishing Co., 1941/1998).

197 **One April morning in 2002:** "Racing Pigeon Returns—Five Years Late," *Manchester Evening News*, May 7, 2005.

197 **The race was held:** J. T. Hagstrum, "Infrasound and the avian navigational map," *J Exp Biol* 203 (2000): 1103–11; J. T. Hagstrum, "Infrasound and the avian navigational map," *J Nav* 54 (2001): 377–91; J. T. Hagstrum, "Atmospheric propagation modeling indicates homing pigeons use loft-specific infrasonic 'map' cues," *J Exp Biol* 216 (2013): 687–99.

198 **The *New York Times* reported:** "The Longest Flight on Record," *New York Times*, August 3, 1885.

198 **A year after the English Channel:** G. Ensley, "Case of the 3,600 disappearing homing pigeons has experts baffled," *Chicago Tribune*, October 18, 1998.

198 **Is it surprising that racing pigeons:** C. Walcott, quoted in G. Ensley, ibid.

199 **The tiny blackpoll warbler:** J. Lathrop, "Tiny songbird discovered to migrate nonstop, 1,500 miles over the Atlantic," news report, University of Massachusetts, Amherst, April 1, 2015.

200 **It's true that the forebrain:** L. N. Voronov et al., "A comparative study of the morphology of forebrain in corvidae in view of their trophic specialization," *Zool Z* 73 (1994): 82–96.

200 **They will accidentally:** W. M. Levi, *The Pigeon* (Sumter, SC: Levi Publishing Co., 1941/1998), 374.

200 **(Although, as one pigeon expert):** Ibid, 374.

200 **And if a bit of nesting material:** "But this is not a fair criticism for the reason that the pigeon's nest is oft-times very neat while the sparrow builds a notoriously untidy one." Ibid.

200 **They're handy with numbers:** D. Scarf et al., "Pigeons on par with primates in numerical competence," *Science* 334 (2011): 1664.

200 **In a laboratory version:** W. T. Herbranson and J. Schroeder, "Are birds smarter than mathematicians? Pigeons (*Columba livia*) perform optimally on a version of the Monty Hall Dilemma," *J Comp Psychol* 124 (2010): 1–13.

201 **(When the Monty Hall Dilemma):** M. vos Savant, "Ask Marilyn," *Parade*, September 9, 1990; December 2, 1990; February 17, 1991; July 7, 1991.

201 **Initially the birds choose:** Walter Herbranson, personal communication, June 4, 2015.

201 **Their successful approach to the problem:** One can solve the problem using either classical probability or empirical probability. In the Monty Hall Dilemma, humans tend to use classical probability. The problem is that we don't use it properly. Pigeons, on the other hand, likely use empirical probability.

201 **a skill the American psychologist:** W. James, *Principles of Psychology*, vol. 1 (New York: Holt, 1890), 459–60.

201 **Alex was not only nearly flawless:** I. M. Pepperberg, "Acquisition of the same/different concept by an African grey parrot (*Psittacus erithacus*): Learning with respect to categories of color, shape, and material," *Anim Learn Behav* 15 (1987): 423–32; Irene Pepperberg, personal communication, May 8, 2015.

202 **Still, pigeons do very well:** M. J. Morgan et al., "Pigeons learn the concept of an 'A,'" *Perception* 5 (1976): 57–66; S. Watanabe, "Discrimination of painting style and beauty: Pigeons use different strategies for different tasks," *Anim Cogn* 14, no. 6 (2011): 797–808; S. Watanabe and S. Masuda, "Integration of auditory and visual information in human face discrimination in pigeons," *Behav Brain Res* 207, no. 1 (2010): 61–69.

202 **They can differentiate between photographs:** R. J. Herrnstein and D. H. Loveland, "Complex visual concept in the pigeon," *Science* 146, no. 3643 (1964): 549–51.

202 **They're highly skilled:** F. A. Soto and W. A. Wasserman, "Asymmetrical interactions in the perception of face identity and emotional expression are not unique to the primate visual system," *J Vision* 11, no. 3 (2011): 24.

202 **They can learn and recall:** J. Fagot and R. G. Cook, "Evidence for large long-term memory capacities in baboons and pigeons and its implications for learning and the evolution of cognition," *PNAS* 103 (2006): 17564–67.

202 **Thanks to breeding:** W. M. Levi, *The Pigeon*, 37.

203 **That's according to the:** Ibid., 1.

203 **"Wherever civilization has flourished":** Ibid.

204 **Cher ami, who:** Ibid., 11.

204 **There was a bird:** Ibid., 10ff.

204 **And Winkie of Scotland:** Ibid., 8.

204 **At its peak in World War II:** Technical Sergeant Clifford Poutre, quoted in *Amarillo Globe Times*, April 1941, http://www.newspapers.com/newspage/29783097/.

204 **Among the most celebrated:** W. M. Levi, *The Pigeon*, 26.

204 **Officials in Cuba:** http://www.cadenagramonte.cu/english/index.php/show/articles/1901:carrier-pigeons-an-alternative-communication-means-at-cuban elections; M. Moore, "China trains army of messenger pigeons," *The Telegraph*, March 2, 2011.

205 **"It is frequently affirmed":** C. Dickens, "Winged Telegraphs," *London Household Word*, February 1850, 454–56

205 **Now we know that's not so:** H. G. Wallraff, "Does pigeon homing depend on stimuli perceived during displacement?" *J Comp Physiol* 139 (1980): 193–201.

205 **True navigation:** The following material on true navigation draws from Richard Holland's excellent synopsis of the current state of the field: R. A. Holland, "True navigation in birds: From quantum physics to global migration," *J Zool* 293 (2014): 1–15.

206 **According to Charles Walcott:** C. Walcott, "Pigeon homing: Observations, experiments and confusions," *J Exp Biol* 199 (1996): 21–27; Charles Walcott quote from report on lecture to Lafayette Racing Pigeon Club, http://www.siegelpigeons.com/news/news-walcott.html.

207 **More than forty years ago, William Keeton:** W. T. Keeton, "Magnets interfere with pigeon homing," *PNAS* 8, no. 1 (1971): 102–6.

207 **The first hint that magnetic fields:** First magnetic field study with European robins: W. Wiltschko and R. Wiltschko, "Magnetic compass of European robins," *Science* 176, no. 4030 (1972): 62–64.

207 **But "sensing magnetic fields":** H. Mouritsen in *Neurosciences: From Molecule to Behavior* (Berlin: Springer Spektrum, 2013), http://link.springer.com/chapter/10.1007/978-3-642-10769-6_20.

207 **One model holds that birds:** M. Zapka et al., "Visual but not trigeminal mediation of magnetic compass information in a migratory bird," *Nature* 461 (2009): 1274–77.

208 **The sensing seems to involve:** Ibid.; M. Liedvogel et al., "Lateralized activation of cluster N in the brains of migratory songbirds," *Eur J Neurosci* 25, no. 4 (2007): 1166–73.

208 **Not long ago, scientists thought:** W. Wiltschko and R. Wiltschko, "Magnetic orientation and magnetoreception in birds and other animals," *J Comp Physiol A* 191 (2005): 675–93; R. Wiltschko and W. Wiltschko, "Magnetoreception," *BioEssays* 28, no. 2 (2006): 157–68; R. Wiltschko et al., "Magnetoreception in birds: Different physical processes for two types of directional responses," *HFSP J* 1, no. 1 (2007): 41–48.

208 **But when researchers looked:** C. D. Treiber et al., "Clusters of iron-rich cells in the upper beaks of pigeons are macrophages not magnetosensitive neurons," *Nature* 484, no. 7394 (2012): 367–70.

208 **New evidence suggests:** R. Wiltschko and W. Wiltschko, "The magnetite-based receptors in the beak of birds and their role in avian navigation," *J Comp Physiol A Neuroethol Sens Neural Behav Physiol* 199 (2013): 89–99; D. Kishkinev et al., "Migratory reed warblers need intact trigeminal nerves to correct for a 1,000 km eastward displacement," *PLoS ONE* 8 (2013): e65847.

208 **Severing the nerve:** D. Kishkinev et al., "Migratory reed warblers need intact trigeminal nerves to correct for a 1,000 km eastward displacement," *PLoS ONE* 8 (2013): e65847.

209 **this time, in tiny balls:** M. Lauwers et al., "An iron-rich organelle in the cuticular plate of avian hair cells," *Curr Biol* 23, no. 10 (2013): 924–29. Every bird, from pigeons to ostriches, has hair cells, each harboring one of these little iron balls. Scientists re-

cently found a group of cells in the brain stems of pigeons that records information on the direction and strength of the magnetic field; the information appeared to be emanating from the bird's inner ear. Perhaps individual neurons in the inner ear detect the direction, intensity, and polarity of magnetic fields and relay this info, providing pigeons with what amounts to a kind of internal GPS.

209 **However, removing the inner:** H. G. Wallraff, "Homing of pigeons after extirpation of their cochleae and lagenae," *Nat New Biol* 236 (1972): 223–24.

209 **In 2014, Mouritsen:** S. Engels et al., "Anthropogenic electromagnetic noise disrupts magnetic compass orientation in a migratory bird," *Nature* 509 (2014): 353–56.

209 **For a long time scientists:** R. Wiltschko and W. Wiltschko, "Avian navigation: From historical to modern concepts," *Anim Behav* 65, no. 2 (2003): 257–72.

209 **The idea goes back to the 1940s:** E. C. Tolman, "Cognitive maps in rats and men," first published in *Psychological Review* 55, no. 4 (1948): 189–208.

210 **(Those who pursued):** T. Lombrozo, "Of rats and men: Edward C. Tolman," *13.7 Cosmos & Culture* (blog), NPR, February 11, 2013, http://www.npr.org/blogs/13.7/2013/02/11/171578224/of-rats-and-men-edward-c-tolman.

210 **Tolman proposed that humans:** E. C. Tolman, "Cognitive maps in rats and men," first published in *Psychological Review* 55, no. 4 (1948): 189–208.

210 **Like rats, it turns out:** R. H. I. Dale, "Spatial memory in pigeons on a four-arm radial maze," *Can J Psychology* 42, no. 1 (1988): 78–83; M. L. Spetch and W. K. Honig, "Characteristics of pigeons' spatial working memory in an open-field task," *Anim Learn Behav* 16 (1988):123–31.

210 **The champs are those:** K. L. Gould et al., "What scatter-hoarding animals have taught us about small-scale navigation," *Philos Trans R Soc Lond B* 365 (2010): 901–14.

211 **Nutcrackers recall the locations:** B. M. Gibson and A. C. Kamil, "The fine-grained spatial abilities of three seed-caching corvids," *Learn Behav* 33, no. 1 (2005): 59–66; A. C. Kamil and K. Cheng, "Way-finding and landmarks: The multiple-bearings hypothesis," *J Exp Biol* 204 (2001): 103–13.

211 **Seven times out of ten:** B. M. Gibson and A. C. Kamil, "The fine-grained spatial abilities of three seed-caching corvids"; D. F. Tomback, "How nutcrackers find their seed stores," *Condor* 82 (1980): 10–19.

211 **One theory holds:** A. C. Kamil and J. E. Jones, "The seed-storing corvid Clark's nutcracker learns geometric relationships among landmarks," *Nature* 390 (1997): 276–79; A. C. Kamil and J. E. Jones, "Geometric rule learning by Clark's nutcrackers (*Nucifraga columbiana*)," *J Exp Psychol Anim Behav Process* 26 (2000): 439–53; P. A. Bednekoff and R. P. Balda, "Clark's nutcracker spatial memory: The importance of large, structural cues," *Behav Proc* 102 (2014): 12–17.

212 **A series of creative experiments:** N. S. Clayton and A. Dickinson, "Episodic-like memory during cache recovery by scrub jays," *Nature* 395 (1998): 272–74; J. M. Dally et al., "The behaviour and evolution of cache protection and pilferage,"*Anim Behav* 72 (2006): 13–23.

212 **Like us, the birds:** Ibid.

212 **To probe whether scrub jays:** C. R. Raby et al., "Planning for the future by western scrub-jays," *Nature* 445, no. 7130 (2007): 919–21.

213 **"Whether jays 'pre-experience'":** L. G. Cheke and N. S. Clayton, "Eurasian jays (*Garrulus glandarius*) overcome their current desires to anticipate two distinct future needs and plan for them appropriately," *Biol Lett* 8 (2012): 171–75.

213 **It relies on spatial:** S. Watanabe and N. S. Clayton, "Observational visuospatial encoding of the cache locations of others by western scrub-jays (*Aphelocoma californica*)," *J Ethol* 25 (2007): 271–79; J. M. Thom and N. S. Clayton, "Re-caching by western scrub-jays (*Aphelocoma californica*) cannot be attributed to stress," *PLoS ONE* 8, no. 1 (2013): e52936.

214 **And they do it, apparently:** S. D. Healy and T. A. Hurly, "Spatial memory in rufous hummingbirds (*Selaphorus rufus*): A field test," *Anim Learn Behav* 23 (1995): 63–68.

214 **a tiny bright orange bird:** Cornell Lab of Ornithology Web site, http://www.allaboutbirds.org/guide/rufous_hummingbird/id.

214 **Healy's recent work:** I. N. Flores-Abreu et al., "One-trial spatial learning: Wild hummingbirds relocate a reward after a single visit," *Anim Cogn* 15, no. 4 (2012): 631–37.

214 **And they can return:** M. Bateson et al., "Context-dependent foraging decisions in rufous hummingbirds," *Proc R Soc B* 270 (2003): 1271–76.

215 **Moreover they keep track:** S. D. Healy, "What hummingbirds can tell us about cognition in the wild," *Comp Cogn Behav* 8 (2013): 13–28.

215 **Healy's research suggests:** New studies suggest the hummingbirds don't use geometry but do use all sorts of subtle visual cues, including landmarks: T. A. Hurly et al., "Wild hummingbirds rely on landmarks not geometry when learning an array of flowers," *Anim Cogn* 17, no. 5 (2014): 1157–65.

215 **But no one had really tested:** N. Blaser et al., "Testing cognitive navigation in unknown territories: Homing pigeons choose different targets," *J Exp Biol* 216, pt. 16 (2013): 3213–31.

217 **In studying brain:** J. O'Keefe and L. Nadel, *The Hippocampus as a Cognitive Map* (Oxford: Oxford University Press, 1978).

217 **New research shows:** J. F. Miller, "Neural activity in human hippocampal formation reveals the spatial context of retrieved memories," *Science* 342 (2013): 1111–14.

217 **A bigger hippocampus:** T. C. Roth et al., "Is bigger always better? A critical appraisal of the use of volumetric analysis in the study of the hippocampus," *Philos Trans R Soc Lond B* 365 (2010): 915–31.

217 **Relative to their whole:** B. J. Ward et al., "Hummingbirds have a greatly enlarged hippocampal formation," *Biol Lett* 8 (2012): 657–59. Ward suggests that other factors could contribute to HF enlargement in hummingbirds—their hovering flight, for instance, which contributes to a "unique brain morphology." It's also possible that the "relative size of the hippocampus in hummingbirds is a result of a reduction in the size of other telencephalic regions" (p. 658).

217 **Brood parasites such:** J. R. Corfield et al., "Brain size and morphology of the brood-parasitic and cerophagous honeyguides (Aves: Piciformes)," *Brain Behav Evol* 81, no. 3 (2013): 170–86.

217 **"This makes sense":** Louis Lefebvre, interviews, February 2012.

218 **Female cowbirds have:** M. F. Guigueno et al., "Female cowbirds have more accurate spatial memory than males," *Biol Lett* 10, no. 2 (2014): 20140026.

218 **Homing pigeons have:** G. Rehkämper et al., "Allometric comparison of brain weight and brain structure volumes in different breeds of the domestic pigeon, *Columba livia* f.d. (fantails, homing pigeons, strassers)," *Brain Behav Evol* 31, no. 3 (1988): 141–49.

218 **Not long ago:** J. Cnotka et al., "Navigational experience affects hippocampus size in homing pigeons," *Brain Behav Evol* 72 (2008): 233–38.

219 **In any case, the size of a pigeon's:** By contrast, research by Vladimir Pravosudov and his team on the hippocampus in food-caching birds "suggests that many brain attributes (e.g., the number of adult neurons) are actually not very plastic and do not change under different conditions," he says. "In other words, it is likely that many of these attributes are heritable and the differences between populations have likely been produced by natural selection acting on memory rather than by individuals adjusting to changing conditions." V. Pravosudov, personal communication, January 2015.

219 **British researchers discovered:** K. Woollett and E. A. Maguire, "Acquiring 'the Knowledge' of London's layout drives structural brain changes," *Curr Biol* 21 (2011): 2109–14.

219 **in what has been deemed:** M. Harris, "Nokia says London is most confusing city," *TechRadar*, November 27, 2008, http://www.techradar.com/us/news/world-of-tech/phone-and-communications/mobile-phones/car-tech/satnav/nokia-says-london-is-most-confusing-city-489141.

219 **The scientists found:** But gaining the Knowledge may come at a cost. The gifted cabbies did poorly on tests of other kinds of spatial memory that involved acquiring or retrieving new visuospatial information. And they had less gray matter volume in their anterior hippocampi.

219 **Indeed, when researchers:** K. Konishi and V. Bohbot, "Spatial navigational strategies correlated with gray matter in the hippocampus of healthy older adults tested in a virtual maze," *Front Aging Neurosci* 5 (2013): 1.

220 **John Huth, a physics:** J. Huth, "Losing our way in the world," *New York Times*, Sunday Review, July 20, 2013.

221 **I was interested:** L. Boroditsky, "Lost in Translation," *Wall Street Journal*, July 23, 2010; L. Boroditsky, "How language shapes thought," *Scientific American*, February 2011.

221 **They carry no map:** A. Michalik et al., "Star compass learning: How long does it take?" *J Ornithol* 155 (2014): 225–34.

221 **After all, dung beetles:** M. Dacke, "Dung beetles use the Milky Way for orientation," *Curr Biol* 23, no. 4 (2013): 298–300.

223 **That the sparrows:** K. Thorup et al., "Evidence for a navigational map stretching across the continental U.S. in a migratory songbird," *PNAS* 104, no. 46 (2007): 18115–19.

223 **The experiment also suggested:** K. Thorup and R. A. Holland, "The bird GPS—long-range navigation in migrants," *J Exp Biol* 212 (2009): 3597–3604. These results confirmed what scientists knew from an impressive experiment carried out on starlings in the 1950s, in which more than eleven thousand starlings caught on migration in the Netherlands were transported to Switzerland. Adult birds were recovered on the way to their normal wintering grounds in the south of England in northwest France. Juveniles were recovered in a more southwesterly direction, say Thorup and Holland, "corresponding to normal direction of migration through the Netherlands."

223 **Inexperienced whoopers shadow:** T. Mueller et al., "Social learning of migratory performance," *Science* 341, no. 6149 (2013): 999–1002. This study found that young birds that followed older birds tended to veer off course almost 40 percent less than those that forayed off on their own. The ability of a whooper to stick to a direct flight path rose steadily each year up to about age five.

224 **We know this because:** K. Thorup and R. A. Holland, "Understanding the migratory orientation program of birds: Extending laboratory studies to study free-flying migrants in a natural setting," *Integ Comp Biol* 50, no. 3 (2010): 315–22.

224 **polarized light cues available at sunset:** Patterns of polarized light also appear to play a key role in navigation. Many of the nocturnal migrants start their flights at sunset or a little after. Birds apparently use the polarized light patterns to provide information on initial migratory flight directions.

224 **In one displacement experiment:** R. Mazzeo, "Homing of the Manx shearwater," *Auk* 70 (1953): 200–201.

225 **To use these gradients:** R. A. Holland, "True navigation in birds: From quantum physics to global migration," *J Zool* 293 (2014): 1–15.

225 **Holland and a colleague:** R. A. Holland and B. Helm, "A strong magnetic pulse affects the precision of departure direction of naturally migrating adult but not juvenile birds," *J R Soc Interface* (2013), doi:10.1098/rsif.2012.1047.

225 **A team led by Nikita Chernetsov:** D. Kishkinev et al., "Migratory reed warblers need intact trigeminal nerves to correct for a 1,000 km eastward displacement," *PLoS ONE* 8, no. 6 (2013): e65847.

226 **According to Jon Hagstrum:** J. T. Hagstrum, "Infrasound and the avian navigational map," *J Exp Biol* 203 (2000): 1103–11; J. T. Hagstrum, "Infrasound and the avian navigational map," *J Nav* 54 (2001): 377–91; J. T. Hagstrum, "Atmospheric propagation modeling indicates homing pigeons use loft-specific infrasonic 'map' cues," *J Exp Biol* 216 (2013): 687–99.

226 **It was April 2014:** H. M. Streby et al., "Tornadic storm avoidance behavior in breeding songbirds," *Curr Biol* (2014), doi:10.1016/j.cub.2014.10.079.

227 "Similar to the way": J. T. Hagstrum, personal communication, January 13, 2014.

227 "The anecdotal evidence": Henrik Mouritsen, personal communication, March 5, 2015.

227 Intrigued by the disappearance: J. T. Hagstrum, "Atmospheric propagation modeling indicates homing pigeons use loft-specific infrasonic 'map' cues," *J Exp Biol* 216 (2013): 687–99.

228 "This is weak evidence": R. A. Holland, "True navigation in birds: From quantum physics to global migration," *J Zool* 293 (2014): 1–15; Richard Holland, personal communication, March 23, 2015.

228 The idea that odor: F. Papi et al., "The influence of olfactory nerve section on the homing capacity of carrier pigeons," *Monit Zool Ital* 5 (1971): 265–67.

228 At around the same time: H. G. Wallraff, "Weitere Volierenversuche mit Brieftauben: Wahrscheinlicher Einfluss dynamischer Faktorender Atmosphare auf die Orientierung," *Z Vgl Physiol* 68 (1970): 182–201.

229 It concerns a fluke: B. L. Finlay and R. B. Darlington, "Linked regularities in the development and evolution of mammalian brains," *Science* 268 (1995): 1578.

229 In nearly all vertebrates: K. E. Yopak et al., "A conserved pattern of brain scaling from sharks to primates," *PNAS* 107, no. 29 (2010): 12946–51.

229 This is true for birds: S. Healy and T. Guilford, "Olfactory bulb size and nocturnality in birds," *Evolution* 44, no. 2 (1990): 339.

230 "An extraordinary development": C. H. Turner, "A few characteristics of the avian brain," *Science XIX*, no. 466 (1892): 16–17.

230 Later, scientists planted electrodes: M. H. Sieck and B. M. Wenzel, "Electrical activity of the olfactory bulb of the pigeon," *Electroenceph Clin Neurophysiol* 26 (1969): 62–69.

230 Blue petrels—seabirds: F. Bonadonna, "Evidence that blue petrel, *Halobaena caerulea*, fledglings can detect and orient to dimethyl sulfide," *J Exp Biol* 209 (2006): 2165–69.

230 These petrels nest: F. Bonadonna, "Could osmotaxis explain the ability of blue petrels to return to their burrows at night?" *J Exp Biol* 204 (2001): 1485–89.

230 Blue tits feeding: L. Amo et al., "Predator odour recognition and avoidance in a songbird," *Funct Ecol* 22 (2008): 289–93.

230 And they'll sniff: A. Mennerat, "Aromatic plants in nests of the blue tit *Cyanistes caeruleus* protect chicks from bacteria," *Oecologia* 161, no. 4 (2009): 849–55.

230 Small seabirds called: S. P. Caro and J. Balthazart, "Pheromones in birds: Myth or reality?" *J Comp Physiol A Neuroethol Sens Neural Behav Physiol* 196, no. 10 (2010): 751–66.

231 Zebra finches, whose bulbs: E. T. Krause et al., "Olfactory kin recognition in a songbird," *Biol Lett* 8, no. 3 (2012): 327–29.

231 A specialist in cognition: L. F. Jacobs, "From chemotaxis to the cognitive map: The function of olfaction," *Proc Natl Acad Sci* 109 (2012): 10693–700.

231 **Anna Gagliardo of the University of Pisa:** A. Gagliardo et al., "Oceanic navigation in Cory's shearwaters: Evidence for a crucial role of olfactory cues for homing after displacement," *J Exp Biol* 216 (2013): 2798–2805.

231 **To find out how:** Ibid.

232 **based on the work of Papi:** F. Papi, *Animal Homing* (London: Chapman & Hall, 1992); H. G. Wallraff, *Avian Navigation: Pigeon Homing as a Paradigm* (Berlin: Springer, 2005).

232 **The first part is a low-resolution:** L. F. Jacobs, "From chemotaxis to the cognitive map : The function of olfaction," *Proc Natl Acad Sci* 109 (2012): 10693–700.

232 **When Wallraff sampled:** H. G. Wallraff and M. O. Andreae, "Spatial gradients in ratios of atmospheric trace gases: A study stimulated by experiments on bird navigation," *Tellus B Chem Phys Meteorol* 52 (2000): 1138–57; H. G. Wallraff, "Ratios among atmospheric trace gases together with winds imply exploitable information for bird navigation: A model elucidating experimental results," *Biogeosciences* 10 (2013): 6929–43.

233 **One study found:** P. E. Jorge et al., "Activation rather than navigational effects of odours on homing of young pigeons," *Curr Biol* 19 (2009): 1–5.

233 **If this study holds:** R. A. Holland, "True navigation in birds: From quantum physics to global migration," *J Zool* 293 (2014): 1–15.

233 **Still, a recent experiment:** R. A. Holland et al., "Testing the role of sensory systems in the migratory heading of a songbird," *J Exp Biol* 212 (2009): 4065–71.

233 **Moreover, when scientists looked:** A. Rastogi et al., "Phase inversion of neural activity in the olfactory and visual systems of a night-migratory bird during migration," *Eur J Neurosci* 34 (2011): 99–109.

234 **In Blaser's study:** N. Blaser et al., "Testing cognitive navigation in unknown territories: Homing pigeons choose different targets," *J Exp Biol* 216, pt. 16 (2013): 3213–31.

234 **A pigeon raised:** C. Walcott, "Multi-modal orientation in homing pigeons," *Integr Comp Bio* 45 (2005): 574–81.

234 **Another pigeon was a champion:** Ibid.

235 **"Humans excel at cognitive":** M. Shanahan, "The brain's connective core and its role in animal cognition," *Philos Trans R Soc Lond B* 367, no. 1603 (2012): 2704–14.

236 **To figure out how:** M. Shanahan et al., "Large-scale network organisation in the avian forebrain: A connectivity matrix and theoretical analysis," *Front Comput Neurosci* 7, no. 89 (2013), doi: 10.3389/fncom.2013.00089.

Eight: Sparrowville: *Adaptive Genius*

240 **In his book:** T. R. Anderson, *Biology of the Ubiquitous House Sparrow* (Oxford: Oxford University Press, 2006), 9.

240 **our avian shadow:** S. Steingraber, "The fall of a sparrow," *Orion Magazine*, 2008.

240 **This is a result:** A. D. Barnosky et al., "Has the earth's sixth mass extinction already arrived?" *Nature* 471 (2011): 51–57.

240 **The habitats that birds:** R. E. Green, "Farming and the fate of wild nature," *Science* 307 (2005): 550. "Farming is now one of the most severe threats faced by the world's birds," says Green. Roughly half the world's surface has been converted to grazed land or cultivated crops. More than half of the world's forests have been lost in that land conversion. Agriculture is the major current and likely future threat to bird species, especially in developing countries.

241 **The ornithologist Pete Dunn:** P. Dunn, *Essential Field Guide Companion* (Boston: Houghton Mifflin, 2006), 679.

241 **Today there are millions:** Story of house sparrow range expansion comes from T. R. Anderson, *Biology of the Ubiquitous House Sparrow* (Oxford: Oxford University Press, 2006), 21–30.

241 **The first sixteen birds:** C. Lever, *Naturalized Birds of the World* (New York: John Wiley, 1987).

241 **In 1889, just a few decades:** E. A. Zimmerman, "House Sparrow History," *Sialis*, http://www.sialis.org/hosphistory.htm.

242 **Now the humble house sparrow:** Partners in Flight Science Committee 2012. Species Assessment Database, version 2012, http://rmbo.org/pifassessment.

242 **Because of its characteristic:** T. R. Anderson, *Biology of the Ubiquitous House Sparrow* (Oxford: Oxford University Press, 2006), 283–84.

242 **When one scientist:** P. A Gowaty, "House sparrows kill eastern bluebirds," *J Field Ornithol* (Summer 1984): 378–80.

242 **Of thirty-nine known house sparrow:** D. Sol et al., "Behavioural flexibility and invasion success in birds," *Anim Behav* 63 (2002): 495–502.

243 **Sol, an ecologist:** D. Sol et al., "The paradox of invasion in birds: Competitive superiority or ecological opportunism?" *Oecologia* 169, no. 2 (2012): 553–64.

243 **Some years ago:** D. Sol and L. Lefebvre, "Behavioural flexibility predicts invasion success in birds introduced to New Zealand," *Oikos* 90 (2000): 599–605.

243 **When Sol later looked:** D. Sol et al., "Unraveling the life history of successful invaders," *Science* 337 (2012): 580.

244 **Successful amphibian and reptile:** Amphibians and reptiles: J. J. Amiel et al., "Smart moves: Effects of relative brain size on establishment success of invasive amphibians and reptiles," *PLoS ONE* 6 (2011): e18277. Mammals: D. Sol et al., "Brain size predicts the success of mammal species introduced into novel environments," *Am Nat* 172 (2008): S63–S71.

244 **For a bird to succeed:** D. Sol et al., "Exploring or avoiding novel food resources? The novelty conflict in an invasive bird," *PLoS ONE* 6, no. 5 (2011): 219535. According to Sol and colleagues, a population of birds "that readily tastes new foods or adopts novel foraging strategies is more pre-adapted to survive and reproduce in a novel environment."

244 **In Normal, Illinois:** J. E. C. Flux and C. F. Thompson, "House sparrows taking insects from car radiators," *Notornis* 33, no. 3 (1986): 190–91.

244 **Sparrows have also:** R. K. Brooke, "House sparrows feeding at night in New York," *Auk* 88 (1971): 924.

244 **A Missouri biologist:** J. L. Tatschl, "Unusual nesting site for house sparrows," *Auk* 85 (1968): 514.

245 **In one weeklong period:** B. D. Bell, "House sparrows collecting feathers from live feral pigeons," *Notornis* 41 (1994): 144–45.

245 **In some cities:** M. Suárez-Rodriguez et al., "Incorporation of cigarette butts into nests reduces nest ectoparasite load in urban birds; new ingredients for an old recipe?" *Biol Lett* 9, no. 1 (2012): 201220921.

245 **When it comes to foraging:** T. Anderson, *Biology of the Ubiquitous House Sparrow* (Oxford: Oxford University Press, 2006), 246–82.

245 **Sparrows have been seen:** K. Rossetti, "House sparrows taking insects from spiders' webs," *British Birds* 76 (1983): 412.

245 **On the Hawaiian island:** H. Kalmus, "Wall clinging: Energy saving of the house sparrow *Passer domesticus*," *Ibis* 126 (1982): 72–74.

246 **Some years ago:** R. Breitwisch and M. Breitwisch, "House sparrows open an automatic door," *Wilson Bulletin* 103 (1991): 4.

246 **According to one account:** R. E. Brockie and B. O'Brien, "House sparrows (*Passer domesticus*) opening autodoors," *Notornis* 51 (2004): 52.

246 **In his book *The Wind Birds*:** P. Matthiessen, *The Wind Birds* (New York: Viking Press, 1973), 20.

247 **When Lynn Martin:** L. B. Martin and L. Fitzgerald, "A taste for novelty in invading house sparrows, *Passer domesticus*," *Behav Ecol* 16, no. 4 (2005): 702–7.

247 **The pair found:** A. Liker and V. Bokony, "Larger groups are more successful in innovative problem solving in house sparrows," *PNAS* 106, no. 19 (2009): 7893–98.

248 **Among Arabian babblers:** Amanda Ridley, personal communication, April 7, 2015.

248 **Studies show that:** P. R. Laughlin et al., "Groups perform better than the best individuals on letters-to-numbers problems: Effects of group size," *J Pers and Soc Psych* 90, no. 4 (2006): 644–51.

248 **Psychologist Steven Pinker:** S. Pinker, "The cognitive niche: Coevolution of intelligence, sociality, and language," *PNAS* 107, suppl. 3 (2010): 8993–99.

248 **which groups come up:** J. Morand-Ferron and J. L. Quinn, "Larger groups of passerines are more efficient problem-solvers in the wild," *PNAS* 108, no. 38 (2011): 15898–903; L. Aplin et al., "Social networks predict patch discovery in a wild population of songbirds," *Proc R Soc B* 279 (2012): 4199–205.

248 **"There is a great tendency":** E. Selous, *Bird Life Glimpses* (London: George Allen, 1905), 79.

248 **But "uniformity of action":** Quoted in M. M. Nice, "Edmund Selous—An Appreciation," *Bird-Banding* 6 (1935): 90–96. Nice draws from E. Selous, *Realities of Bird Life*

(London: Constable & Co., 1927), 152; E. Selous, *The Bird Watcher in the Shetlands* (London: J. M. Dent & Co., 1905), 232.

249 **how they respond to oxytocin-like:** A. M. Kelly and J. L. Goodson, "Personality is tightly coupled to vasopressin-oxytocin neuron activity in a gregarious finch," *Front Behav Neurosci* 8, no. 55 (2014), doi: 10.3389/fnbeh.2014.0005.

249 **For instance, John Cockrem:** J. F. Cockrem, "Corticosterone responses and personality in birds: Individual variation and the ability to cope with environmental changes due to climate change," *Gen Comp Endocrinol* 190 (2013): 156–63.

249 **Lynn Martin has caught:** A. W. Schrey et al., "Range expansion of house sparrows (*Passer domesticus*) in Kenya: Evidence of genetic admixture and human-mediated dispersal," *J Heredity* 105 (2014): 60–69.

249 **The birds were first:** Lynn Martin, personal communication, March 6, 2015.

249 **Now they're common:** J. D. Parker et al., "Are invasive species performing better in their new ranges?" *Ecology* 94 (2013): 985–94.

249 **Birds farthest from Mombasa:** L. B. Martin et al., "Surveillance for microbes and range expansion in house sparrows," *Proc R Soc B* 281, no. 1774 (2014): 20132690.

249 **The scientists suggest that the stress:** A. L. Liebl and L. B. Martin, "Exploratory behavior and stressor hyper-responsiveness facilitate range expansion of an introduced songbird," *Proc R Soc B* (2012), doi:10.1098/rspb.2012.1606.

250 **When Martin's graduate student:** A. L. Liebl and L. B. Martin, "Living on the edge: Range edge birds consume novel foods sooner than established ones," *Behav Ecol* 25, no. 5 (2014): 1089–96.

250 **In contrast, the leading birds:** This jived with what Martin found in an earlier study comparing two groups of New World house sparrows. The first group, from the city of Colon, Panama, were newcomers. They had been introduced to the country only 30 years earlier, and were actively spreading across it. The other population was a group of staid "old-timers" that had lived as residents of Princeton, New Jersey, for more than 150 years. Martin kept both groups in captivity under similar conditions and then tested their responses to such novel foods as slices of kiwi and finely crushed Life Savers. The birds from Panama happily consumed the new foods, while the New Jersey birds rejected them. See L. B. Martin and L. Fitzgerald, "A taste for novelty in invading house sparrows," *Behav Ecol* 16 (2005): 702–7.

250 **But one bird:** M. J. Afemian et al., "First evidence of elasmobranch predation by a waterbird: Stingray attack and consumption by the great blue heron," *Waterbirds* 34, no. 1 (2011): 117–20.

251 **A brown pelican:** D. L. Bostic and R. C. Banks, "A record of stingray predation by the brown pelican," *Condor* 68, no. 5 (1966): 515–16.

251 **One kea in the alpine:** B. D. Gartell and C. Reid, "Death by chocolate: A fatal problem for an inquisitive wild parrot," *New Zealand Vet J* 55, no. 3 (2007): 149–51.

251 **But, as Lynn Martin:** Lynn Martin, personal communication, March 5, 2015.

251 **Once the birds:** According to Martin and his colleague, "Therefore, selection should

reduce flexibility in individuals residing in environmentally stable areas but favor it in novel and/or variable environments. . . . As flexibility can incur costs, it might not be a viable strategy for all individuals, especially those that persist at sites far from range edges, and therefore selection should begin to hone phenotypes to match local conditions." L. B. Martin and L. Fitzgerald, "A taste for novelty in invading house sparrows," *Behav Ecol* 16 (2005): 702–7.

252 **a penchant for hanging out:** It should be noted, however, that there is no empirical evidence to support the suggestion that sociality is an important feature of successful invaders, as Daniel Sol points out. "The reason is that almost all species that have been introduced are social, perhaps because they are easier to capture or because they are more frequent near human settlements. Thus, the prediction cannot be properly tested." Daniel Sol, personal communication, January 2015.

252 **(The latter, called a bet-hedging strategy):** Daniel Sol, personal communication, April 2015.

252 **(In Toronto, for instance):** R. Johns, "Building owners in new lawsuit over bird collision deaths," American Bird Conservancy media release, 2012, http://www.abcbirds .org/newsandreports/releases/120413.html.

252 **Daniel Sol and his colleagues:** Daniel Sol, personal communication, April 2015; D. Sol et al., "Urbanisation tolerance and the loss of avian diversity," *Ecol Lett* 17, no. 8 (2014): 942–50.

253 **Canadian researchers recently found:** D. S. Proppe et al., "Flexibility in animal signals facilitates adaptation to rapidly changing environments," *PLoS ONE* (2011), doi:10.1371/journal.pone.0025413.

253 **When scientists looked at trends:** S. Shultz, "Brain size and resource specialization predict long-term population trends in British birds," *Proc R Soc B* 272, no. 1578 (2005): 2305–11.

253 **New insights from the farms:** L. O. Frishkoff, "Loss of avian phylogenetic diversity in neotropical agricultural systems," *Science* 345, no. 6202 (2014): 1343–46.

254 **Research by Daniel Sol:** D. Sol et al., "Behavioral drive or behavioral inhibition in evolution: Subspecific diversification in Holarctic passerines," *Evolution* 59, no. 12 (2005): 2669–77; D. Sol and T. D. Price, "Brain size and the diversification of body size in birds," *Am Nat* 172, no. 2 (2008): 170–77.

255 **In early 2014:** B. G. Freeman and A. M. Class Freeman, "Rapid upslope shifts in New Guinean birds illustrate strong distributional responses of tropical montane species to global warming," *PNAS* 111 (2014): 4490–94.

255 **"I find it astonishing":** Ben Freeman, personal communication, February 5, 2015.

256 **I once saw a global map:** P. Kareiva et al., "Conservation in the Anthropocene," *The Breakthrough* (Winter 2012), http://thebreakthrough.org/index.php/journal/past -issues/issue-2/conservation-in-the-anthropocene.

257 **According to projections:** S. Nash, *Virginia Climate Fever* (Charlottesville: University of Virginia Press, 2014), 24.

257 **Great tits, known for:** O. Vedder et al., "Quantitative assessment of the importance of phenotypic plasticity in adaptation to climate change in wild bird populations," *PLoS Biol* (2013), doi:10.1371/journal.pbio.1001605.

258 **These birds have longer:** However, as Daniel Sol points out, "other studies show the contrary: a longer generation time increases the response to climate change." See B.-E. Saether, "Climate driven dynamics of bird populations: Processes and patterns," *BOU Proceedings—Climate Change and Birds* (2010).

258 **If warming changes:** S. Shultz, "Brain size and resource specialization predict long-term population trends in British birds," *Proc R Soc B* 272, no. 1578 (2005): 2305–11; D. Sol et al., "Big brains, enhanced cognition and response of birds to novel environments," *PNAS* 102 (2005): 5460–65.

258 **Since the 1980s:** A. J. Baker, "Rapid population decline in red knots: fitness consequences of decreased refuelling rates and late arrival in Delaware Bay," *Proc Roy Soc B* 271 (2004): 875–82.

258 **Shifting temperatures may:** H. Galbraith et al., "Predicting vulnerabilities of North American shorebirds to climate change," *PLoS ONE* (2014), doi:10.1371/journal.pone.0108899.

259 **Its habitat is expected:** http://climate.audubon.org/birds/mouchi/mountain-chickadee.

259 **Moreover, global warming:** C. A. Freas et al., "Elevation-related differences in memory and the hippocampus in mountain chickadees, *Poecile gambeli*," *Anim Behav* 84 (2012): 121–27.

259 **According to Vladimir Pravosudov:** Vladimir Pravosudov, personal communication, January 29, 2015.

259 **"That's the lowest total ever":** Ben Freeman, personal communication, February 26, 2015.

259 **Indeed, around the globe:** G. De Coster et al., "Citizen science in action—evidence for long-term, region-wide house sparrow declines in Flanders, Belgium," *Landscape Urban Plan* 134 (2015): 139–46; L. M. Shaw et al., "The house sparrow *Passer domesticus* in urban areas—reviewing a possible link between post-decline distribution and human socioeconomic status," *J Ornithol* 149, no. 3 (2008): 293–99.

259 **Its decline generates:** http://www.rspb.org.uk/discoverandenjoynature/discoverandlearn/birdguide/redliststory.aspx.

259 **The survival of nestlings:** W. J. Peach et al., "Reproductive success of house sparrows along an urban gradient," *Anim Conserv* 11, no. 6 (2008): 493–503; http://www.rspb.org.uk/news/details.aspx?id=tcm:9-203663; D. Adam, "Leylandii may be to blame for house sparrow decline, say scientists," *Guardian*, 2008, http://www.theguardian.com/environment/2008/nov/20/wildlife-endangeredspecies.

259 **Gardens converted to parking lots:** G. Seress, "Urbanization, nestling growth and reproductive success in a moderately declining house sparrow population," *J Avian Biol* 43 (2012): 403–14.

259 **Some evidence from Israel:** Y. Yom-Tov, "Global warming and body mass decline in Israeli passerine birds," *Proc R Soc B* 268 (2001): 947–52.

259 **Lynn Martin says he's skeptical:** Lynn Martin, personal communication, March 5, 2015.

260 **The final paragraph:** T. R. Anderson, *The Biology of the Ubiquitous House Sparrow* (Oxford: Oxford University Press, 2006), 437.

260 **Scientists are still turning:** P. C. Rasmussen et al., "Vocal divergence and new species in the Philippine hawk owl *Ninox philippensis* complex," *Forktail* 28 (2012): 1–20; J. B. C. Harris, "New species of *Muscicapa* flycatcher from Sulawesi, Indonesia," *PLoS ONE* 9, no. 11 (2014): e112657; P. Alström et al., "Integrative taxonomy of the russet bush warbler *Locustella mandelli* complex reveals a new species from central China," *Avian Res* 6, no. 1 (2015), doi:10.1186/s40657-015-0016-z.

260 **A new study suggests that crows:** A. Smirnova et al., "Crows spontaneously exhibit analogical reasoning," *Curr Biol* (2014), doi:http://dx.doi.org/10.1016/j.cub.2014.11.063.

261 **According to Richard Prum:** R. O. Prum, "Coevolutionary aesthetics in human and biotic artworlds," *Biol Philos* 28, no. 5 (2013): 811–32.

261 **As the ornithologist Richard F. Johnston:** R. F. Johnston, quoted in T. R. Anderson, *The Biology of the Ubiquitous House Sparrow* (Oxford: Oxford University Press, 2006), 31.

262 **"If you evolved":** Gavin Hunt, personal communication, January 2015.

262 **But as one of the scientists:** L. O. Frishkoff, "Loss of avian phylogenetic diversity in neotropical agricultural systems," *Science* 345, no. 6202 (2014): 1343–46.

262 **A new study comparing the genomes:** M. N. Romanov et al., "Reconstruction of gross avian genome structure, organization and evolution suggests that the chicken lineage most closely resembles the dinosaur avian ancestor," *BMC Genomics* 15, no. 1 (2014): 1060.

262 **Arthur Cleveland Bent:** A. C. Bent, *Life Histories of North American Gallinaceous Birds* (Washington, DC: U.S. Government Printing Office, 1932), 335.

263 **As Aldo Leopold reminds:** A. Leopold, *A Sand County Almanac* (London: Oxford University Press, 1966), 137.

263 **Evidence suggests that the "big bang":** E. D. Jarvis et al., "Whole-genome analyses resolve early branches in the tree of life of modern birds," *Science* 346, no. 6215 (2014): 1321–31.

264 **"As a human being":** A. Einstein in a letter to Queen Elisabeth of Belgium, September 19, 1932.

264 **"Measuring the fitness benefits":** S. D. Healy, "Animal cognition: The tradeoff to being smart," *Curr Biol* 22, no. 19 (2012): R840–41.

264 **Daniel Sol has data:** Daniel Sol, personal communication, January 2015.

265 **One study of wild great tits:** L. Cauchard et al., "Problem-solving performance is

correlated with reproductive success in a wild bird population," *Anim Behav* 85 (2013): 19–26. Cauchard and her colleagues presented breeding pairs of tits with a tricky problem-solving task and then correlated the parents' performance with their reproductive success. The team built nesting boxes with a kind of trapdoor that could be opened only by pulling a string. Nests where at least one parent solved the task had higher nestling survival than nests where both parents couldn't solve the task.

265 **However, in a close look:** E. Cole et al., "Cognitive ability influences reproductive life history variation in the wild," *Curr Biol* 22 (2012): 1808–12.

265 **(The same has been found):** D. Y. Kozlovsky et al., "Elevation-related differences in parental risk-taking behavior are associated with cognitive variation in mountain chickadees," *Ethology* 121, no. 4 (2015): 383–94; Vladimir Pravosudov, personal communication, January 25, 2015.

265 **"Were the good problem solvers":** Neeltje Boogert, personal communication, April 2015.

266 **On the island of Barbados:** Simon Ducatez, interview, February 2012; S. Ducatez, "Problem-solving and learning in Carib grackles: Individuals show a consistent speed-accuracy tradeoff," *Anim Cogn* 18, no. 2 (2015): 485–96.

266 **"Bolder individuals tend":** Daniel Sol, personal communication, January 2015.

INDEX